# NATO ASI Series

## Advanced Science Institutes Series

*A series presenting the results of activities sponsored by the NATO Science Committee, which aims at the dissemination of advanced scientific and technological knowledge, with a view to strengthening links between scientific communities.*

The Series is published by an international board of publishers in conjunction with the NATO Scientific Affairs Division

| | |
|---|---|
| A  Life Sciences | Plenum Publishing Corporation |
| B  Physics | London and New York |
| C  Mathematical and Physical Sciences | Kluwer Academic Publishers |
| D  Behavioural and Social Sciences | Dordrecht, Boston and London |
| E  Applied Sciences | |
| F  Computer and Systems Sciences | Springer-Verlag |
| G  Ecological Sciences | Berlin Heidelberg New York |
| H  Cell Biology | London Paris Tokyo Hong Kong |
| I  Global Environmental Change | Barcelona Budapest |

### PARTNERSHIP SUB-SERIES

| | |
|---|---|
| 1. Disarmament Technologies | Kluwer Academic Publishers |
| 2. Environment | Springer-Verlag/Kluwer Academic Publishers |
| 3. High Technology | Kluwer Academic Publishers |
| 4. Science and Technology Policy | Kluwer Academic Publishers |
| 5. Computer Networking | Kluwer Academic Publishers |

*The Partnership Sub-Series incorporates activities undertaken in collaboration with NATO's Cooperation Partners, the countries of the CIS and Central and Eastern Europe, in Priority Areas of concern to those countries.*

## NATO-PCO DATABASE

The electronic index to the NATO ASI Series provides full bibliographical references (with keywords and/or abstracts) to about 50 000 contributions from international scientists published in all sections of the NATO ASI Series. Access to the NATO-PCO DATABASE compiled by the NATO Publication Coordination Office is possible in two ways:

- via online FILE 128 (NATO-PCO DATABASE) hosted by ESRIN,
  Via Galileo Galilei, I-00044 Frascati, Italy.

- via CD-ROM "NATO Science & Technology Disk" with user-friendly retrieval software in English, French and German (© WTV GmbH and DATAWARE Technologies Inc. 1992).

The CD-ROM can be ordered through any member of the Board of Publishers or through NATO-PCO, Overijse, Belgium.

Series H: Cell Biology, Vol. 97

Springer
Berlin
Heidelberg
New York
Barcelona
Budapest
Hong Kong
London
Milan
Paris
Santa Clara
Singapore
Tokyo

# Post-transcriptional Control
# of Gene Expression

Edited by

## Orna Resnekov

Harvard University, Biological Laboratories
16 Divinity Avenue
Cambridge, MA 02138, USA

## Alexander von Gabain

Vienna Biocenter, Department of Microbiology and Genetics
Dr. Bohr-Gasse 9, A-1030 Vienna, Austria

Springer

Published in cooperation with NATO Scientific Affairs Division

Proceedings of the NATO Advanced Study Institute "Post-transcriptional
Control of Gene Expression", held on the island of Spetsai, Greece,
August 3–14,1992

Library of Congress Cataloging-in-Publication Data


Post-transcriptional control of gene expression / edited by Orna
    Resnekov, Alexander von Gabain.
        p.    cm. -- (NATO ASI series.  Series H, Cell biology ; vol.
    97)
      "Proceedings of the NATO Advanced Study Institute "Post
    -transcriptional Control of Gene Expression," held on the island of
    Spetsai, Greece, August, 1992"--T.p. verso.
      Includes bibliographical references and index.
    **Additional material to this book can be downloaded from http://extra.springer.com.**

    ISBN-13:978-3-642-64609-6
      1. Genetic transcription--Regulation--Congresses.  2. Genetic
    regulation--Congresses.   I. Resnekov, Orna, 1958-   . II. Gabain,
    Alexander von, 1950-   . III. Series.
    QH450.2.P67   1996
    574.87 3223--dc20                                          96-14087
                                                                    CIP

ISBN-13:978-3-642-64609-6          e-ISBN-13:978-3-642-60929-9
DOI:10.1007/978-3-642-60929-9

© Springer-Verlag Berlin Heidelberg 1996
Softcover reprint of the hardcover 1st edition 1996

Typesetting: Camera ready by authors/editors
Printed on acid-free paper
SPIN: 10062127        31/3137 - 5 4 3 2 1 0

## Preface

In August of 1992 about 110 students and 18 lecturers gathered on the island of Spetsai in Greece for an 11 day NATO/FEBS Advanced Studies Institute on the topic "Post-transcriptional Control of Gene Expression". Lectures were presented on the following five topics:

1) The general principles of translation and mRNA stability
2) Protein folding, transport and modification
3) Antisense RNA
4) Translational control, regulation, frameshifting and fidelity
5) mRNA stability, processing and modification

In addition, tutorials (in which the students and lecturers met in small informal groups) were held to discuss questions that arose from the lectures and to compare and contrast different types of post-transcriptional control.

During the meeting the lecturers were invited to contribute manuscripts (on their own area of interest) to a book that would be published subsequently, and the following volume is the culmination of that effort. As will be readily apparent, some lecturers chose to combine their efforts on a topic in one chapter, while in other cases chapters were best covered by a single lecturer. We have tried to avoid grouping the chapters into sections, hopefully encouraging the reader to explore the scope of the presentations.

To ensure that a representative selection of the topics presented at the Advanced Studies Institute were covered in this volume, we have been permissive with the format and date of submission of the manuscripts. We accept full responsibility for any failings of the book, including the fact that it will appear in print at a later date than we first anticipated (primarily because of editorial delays). At the end of this section we have compiled an updated list of references to supplement those present in the chapters. We have also included an updated list of citations for references within the chapters that are listed as in press. Nevertheless, we are very happy with the outcome of this book, and very grateful to all the authors for their marvelous contributions.

Orna Resnekov and Alexander von Gabain

## Acknowledgments

We are grateful to NATO and FEBS for their generous support of our Advanced Studies Institute.

We would also like to acknowledge the support of the following sponsors:

Sorin Biomedica S.P.A., Saluggia, Italy
CIBA-Geigy, Basel, Switzerland
Longa AG, Visp, Switzerland
General Secretariat of Research and Development of Greece
Kabi Pharmacia, Stockholm, Sweden
Repligen-Sandoz Research Corporation, Lexington, USA
ASTRA Research Center, India
Svenska Handelsbanken, Stockholm, Sweden

Finally, we gratefully acknowledge the support of our respective research institutions during the completion of this project:

Karolinska Institute, Stockholm , Sweden (A. v. G).

Harvard Biological Laboratories (Losick laboratory)
and the Bunting Institute, Cambridge, Ma. (O.R.).

**Updated list of references:**

Altmann, M.and Trachsel, H. (1994) The yeast *Saccharomyces cerevisiae* system: a powerful tool to study the mechanism of protein synthesis initiation in eukaryotes. Biochimie. **76**:853-61

Beelman, C.A. and Parker, R. (1995) Degradation of mRNA in eukaryotes. Cell. **81**:179-83

Bilger, A., Fox, C.A.,Wahle, E. and Wickens, M. (1994) Nuclear polyadenylation factors recognize cytoplasmic polyadenylation elements. Genes and Development. **8**:1106-16

Blight, M.A., Chervaux, C. and Holland, I.B. (1994) Protein secretion pathway in *Escherichia coli*. Curr. Opin. Biotechnol. **5**:468-74

Carposis, A.J., Houwe, G., Ehretsmann, C.P. and Krish, H.M. (1994) Copurification of *E. coli* RNase E and PNPase: evidence for a specific association between two enzymes important in RNA processing and degradation. Cell. **76**:1-20

Chen, C.Y. and Shyu, A.B. (1994) Selective degradation of early-response-gene mRNAs: functional analyses of sequence features of the AU-rich elements. Mol. Cell. Biol. **14**:8471-82

Clarke, P.A., Pe'ery, T., Ma, Y. and Mathews, M.B. (1994) Structural features of adenovirus 2 virus-associated RNA required for binding to the protein kinase DAI. Nucleic. Acids. Res. **22**:4364-74

Cohen, S.N. (1995) Surprises at the 3' end of prokaryotic RNA. Cell. **80**:829-32

Dehlin, E., von Gabain, A., Alm, G., Dingelmayer, R. and Resnekov, O. (1996) Repression of β-IFN gene expression in virus infected cells is correlated with poly(A) tail elongation. Mol. Cell. Biol. **16**:468-474

Delihas, N. (1995) Regulation of gene expression by trans-coded antisense RNAs. Mol. Microbiol. **15**:411-4

Engelberg-Kulka, H. and Schoulaker-Schwarz, R. (1994) Regulatory implications of translational frameshifting in cellular gene expression. Mol. Microbiol. **11**:3-8

Gray, N.K. and Hentze, M.W. (1994) Regulation of protein synthesis by mRNA structure. Mol. Biol. Rep. **19**:195-200

Hartl, F.U. (1995) Principles of chaperone mediated protein folding. Philos. Trans. R. Soc. Lond. Bio. Sci. **348**:107-112

Hartl, F.U., and Martin, J. (1995) Molecular chaperones in cellular protein folding. Curr. Opin. Struct. Biol. **5**:92-102

Hinnebusch, A.G. (1994) Translational control of GCN4: an *in vivo* barometer of initiation factor activity. Trends Biochem. Sci. **19**:409-14

Jain, C. and Belasco, J. (1995) RNase E autoregulates its synthesis by controlling the degradation rate of its own mRNA in *Escherichia coli*: unusual sensitivity of the *rne* transcript to RNase E activity. Genes and Development, **9**:84-96

Kaufman, R.J. (1994) Control of gene expression at the level of translation initiation. Curr. Opin. Biotechnol. **5**:550-7

Kurland, C.G. (1992) Translational accuracy and the fitness of bacteria. Ann. Rev. Genetics. **26**:29-50

Kutay, U., Ahnert-Hilger, G., Hartman, E., Wiedenmann, B. and Rapoport, T.A. (1995) Transport route for synaptobrevin via a novel pathway of insertion into the endoplasmic reticulum membrane. EMBO J. **14**:217-23

Lundberg, U., Melefors, O., Sohlberg, B., Georgellis, D., and von Gabain, A. (1995) RNase K: one less letter in the alphabet soup. Mol. Microbiol., **17**:595-96

Ma, C.K., Kolesnikow, T., Rayner, J.C., Simons, E.L., Yim, H. and Simons, R.W. (1994) Control of translation by mRNA secondary structure; the importance of the kinetics of structure formation. Mol. Microbiol. **14**:1033-47

McCarthy, J.E. and Brimacombe, R. (1994) Prokaryotic translation: the interactive pathway leading to initiation. Trends Genet. **10**:402-7

McDowall, K.J., Kaberdin, V.R., Wu, S.W., Cohen, S.N., Lin-Chao, S. (1995) Site-specific RNase E cleavage of oligonucleotides and inhibition by stem loops. Nature. **374**:287-90

Nordstrom, K. and Wagner, E.G. (1994) Kinetic aspects of the control of plasmid replication by antisense RNA. Trends Biochem. Sci. **19**:294-300

Pantopoulos, K. and Hentze, M.W. (1995) Rapid responses to oxidative stress mediated by iron regulatory protein. EMBO J. **14**:2917-24

Panzner, S., Dreier, L., Hartmann, E., Kostka, S. and Rapoport, T.A. (1995) Posttranslational protein transport in yeast reconstituted with a purified complex of Sec proteins and Kar2p. Cell. **81**:561-70

Seufert, W., Futcher, B. and Jentsch, S. (1995) Role of a ubiquitin-conjugating enzyme in degradation of S- and M- phase cyclins. Nature. **373**:78-81

Stripecke, R., Olveira, C.C., McCarthy, J.E. and Hentze, M.W. (1994) Proteins binding to 5' untranslated region sites: a general mechanism for translational regulation of mRNAs in human and yeast cells. Mol. Cell. Biol. **14**:5898-909

von-Ahsen, U. and Noller, H.F. (1995) Identification of bases in 16S rRNA essential for tRNA binding at the 30S ribosomal P site. Science. **267**:234-7

Wagner, E.G. and Simons, R.W. (1994) Antisense RNA control in bacteria, phages and plasmids. Ann. Rev. Microbiol. **48**:713-42

Wang, M. and Cohen, S. (1994) ard-1: a human gene that reverses the effects of temperature sensitive deletion mutants in the Escherichia coli *rne* gene and encodes an activity producing RNase E-like cleavages. Proc. Natl. Acad. Sci. USA. **91**:10591-10595

Wennborg, A., Sohlberg, B., Angerer, D., Klein, G. and von Gabain, A. (1995) A human RNase E-like activity that cleaves RNA sequences involved in mRNA stability control. Proc. Natl. Acad. Sci. USA. **92**:7322-6

Wickner, R.B. (1993) Host control of yeast dsRNA virus propagation and expression. Trends Microbiol. **1**:294-9

Xu, F. and Cohen, S. N. (1995) RNA degradation in *Escherichia coli* regulated by 3' adenylation and 5' phosphorylation. Nature. **374**:180-3

Zhang, S., Ruiz-Echevarria, M.J., Quan, Y. and Peltz, S.W. (1995) Identification and characterization of a sequence motif involved in nonsense-mediated mRNA decay. Mol. Cell. Biol. **15**:2231-44

# REFERENCES LISTED AS IN PRESS

## Chapter I

Peltz, S. W., Brown, A. H. and Jacobsen, A. 1993. mRNA destabilization triggered by premature translational termination depends on at least three *cis*-acting sequence elements and one *trans*-acting factor. *Genes and Devel.* 7(9): 1737-54.

## Chapter II

von Gabain, A. Data not published.

Zilhao, R., Camelo, L. and Arraiano, C. M. 1993. DNA sequencing and expression of the gene *rub* encoding *Escherichia coli* ribonuclease II. *Mol Microbiol.* 8(1): 43-51.

Mc Dowall, K. J., Hernandez, R. G., Lin-Chao, S. and Cohen, S. N. 1993. The *ams-1* and *rne 3071* temperature sensitive mutations in the *ams* gene are in close proximity to each other and cause substitution within a domain that resembles a product of the *Escherichia coli mre* locus. *J. Bact.* 175(13): 4245-9

Sanson, B. and Uzan, M. 1993. Dual role of the sequence specific bacteriophage T4 endoribonuclease Reg B. mRNA inactivation and mRNAdestabilization.*J. Mol. Biol.* 233(3): 429-46.

## Chapter III

Hansen, M. J., Schoenberg, M. L. and Belasco, J. G. 1994. The *ompA* 5′ untranslated region impedes a major pathway for mRNA degradation in *Escherichia coli. Mol. Microbiol.* 12(5): 707-16.

## Chapter IV

Schiavi, S. C., Wellington, C. L., Shyu, A.-B., Chen, C.-Y. A., Greenberg, M. E. and Belasco, J. G. 1994. Multiple elements in *c-fos* protein coding region facilitate mRNA deadenylation and decay by a mechanism coupled to translation. *J. Biol. Chem..* 269(5): 3441-8.

Wahle, E. and Keller, W. RNA-protein interactions in mRNA 3′ end formation. *Mol. Biol. Rep.* 18(2): 157-61.

## Chapter V

Guarlezi, C. O. and Pon, C. L. 1993. mRNA-ribosome interaction during initiation of protein synthesis. In: Zimmermann, R. A. and Dahlberg, A. E. (eds) Ribosomal RNA. CRC Press, Inc.

**La Teana, A., Pon, C. L. and Guarlezi, C. O.** 1993. Translation of mRNAs with degenarate initiation triplet AUU displays high IF2 dependence and is subject to IF3 repression. *Proc. Natl. Acad. Sci. USA* . **90**(9): 4161-5.

**Philippe, C., Eyermann, F., Bernard, L., Portier, C., Ehresmann, B. and Ehresmann, C.** 1993. Ribosomal protein S15 from *E. coli* modulates its own translation by trapping the ribosome on the mRNA initiation loading site. *Proc. Natl. Acad. Sci. USA* . **90**(10): 4394-8.

## Chapter VII

**Brimacombe, R., Mitchell, P., Osswald, M., Stade, K. and Bochkariov, D.** 1993. Clustering of modified nucleotides at the functional center of bacterial ribosomal RNA. *FASEB. J.* **7**(1): 161-7.

## Chapter IX

**Draper, D. E.** 1993. Mechanisms of translational initiation and repression in Prokaryotes. In:" Nierhaus K(eds) The translational apparatus, Plenum, New York.

**Laing, L. G. and Draper, D. E.** 1994. Thermodynamics of RNA folding in a conserved ribosomal RNA domain. *J. Mol. Biol.* **237**(5): 560-76.

**Spedding, G. S., Draper, D. E.** 1993. Allosteric mechanism for translational repression in the *E. coli* α operon. *Proc. Natl. Acad. Sci. USA* . **90**(10): 4399-403.

**Spedding, G. S., Gluick, T. C. and Draper, D. E.** 1993. Ribosome initiation complex formation with the pseudoknotted α operon messenger RNA. *J. Mol. Biol.* **229**(3): 609-22.

## Chapter X

**Oliveira, C. C., van der Heuvel, J. J. and McCarthy, J. E. G.** 1993. Inhibition of translational initiation in *Saccharomyces cerevisiae* by secondary structure: the role of the stability and the position of stem-loops inthe mRNA leader. *Molec. Microbiol.* **9**(3): 521-32.

**Rex, G., Surin, B., Besse, G., Schneppe, B. and McCarthy, J. E. G.** 1994. The mechanism of translational coupling in *Escherichia. coli*:: higher-order structure in the *atpHA* mRNA acts as a conformational switch regulating the access of the *de novo* initiating ribosomes. J. *Mol. Chem.* **269**(27): 18118-27.

**Vega Laso, M. R., Zhu, D., Sagliocco, F., Brown, A. J. P., Tuite, M. F. and McCarthy, J. E. G.** 1993. Inhibition of translational initiation in th yeast *Saccharomyces cerevisiae* as a function of the stability and position of hairpin structures in the mRNA leader. *J. Biol. Chem.* **268**(9): 6453-62.

## Chapter XI

Melefors, Ö.and Hentze, M. W. 1993. Translational regulation by mRNA/protein interactions in eukaryotic cells: ferritin and beyond. *Bioessays* **15**(2): 85-90..

Melefors, Ö., Goossen, B., Johansson, H. E., Stripecke, R., Gray, N. K. and Hentze, M. W. 1993. Translational control of 5-aminoalevulinate synthase mRNA by iron responsive elements in erythroid cells. *J. Biol. Chem.* **268**(8): 5974-8.

Gray, N. K., Quick, S., Goossen, B., Constable, A., Hirling, H., Kuhn, L. C. and Hentze, M. W. 1993. Recombinant iron regulatory factor (IRF) functions as an iron responsive element binding protein, a translational repressor and an aconitase. A functional assay for translational repression and direct demonstration of the iron switch. *Eur. J. Biochem.* **218**(2): 657-67.

## Chapter XII

Benhar, I. and Engelberg-Kulka, H. 1993. Frameshifting in the expression of the *Escherichia coli trpR* gene occurs by the bypassing of a segment of its own coding sequence. *J. Bact.* **175**(10): 3204-7.

## Chapter XVI

Jungmann, J., Reins, H. -A., Schobert, C. and Jentsch, S. 1993. Resistance to cadmium mediated by ubiquitin-dependent proteolysis. *Nature* **361**(6410): 369-71..

Zhen, M., Heinlein, R., Jones, D., Jentsch, S. and Candido, E. P. M. 1993. The *ubc-2* gene of *Caenorhabditis elegans* encodes a ubiquitin-conjugating enzyme involved in selective protein degradation. *Mol. Cell. Biol.* **13**(3): 1371-7.

## Chapter XVIII

High, S., Martoglio, B., Gorlich, D., Andersen, S. S. L., Ashford, A. J., Giner, A., Hartmann, E., Prehn, S., Rapoport, T. A., Dobberstein, B. and Brunner, J. 1993. Site specific photocrosslinking reveals that Sec61p and TRAM contact different regions of a membrane inserted signal sequence. *J. Biol. Chem.* **268**(35): 26745-51.

# Table of Contents

# REGULATION OF EUKARYOTIC GENE EXPRESSION AT THE LEVEL OF mRNA STABILITY: EMERGENCE OF GENERAL PRINCIPLES

Stuart W. Peltz[1] and Allan Jacobson[2]
[1]Department of Molecular Genetics and Microbiology
University of Medicine and Dentistry of New Jersey
Robert Wood Johnson Medical School
Piscataway, NJ 08854
and
[2]Department of Molecular Genetics and Microbiology
University of Massachusetts Medical School
Worcester, Massachusetts 01655

## Introduction

To a first approximation, changes in the expression of specific genes are manifested by changes in the steady-state levels of individual mRNAs. Consequently, the regulation of mRNA stability must be an important step in the control of gene expression. In a variety of experimental systems, characterization of the mechanisms involved in the turnover of individual mRNAs has begun. Such studies have led to the identification of essential cis-acting mRNA destabilizing and stabilizing sequences and specific trans-acting factors as well as to the recognition that mRNA turnover requires the involvement of the translation apparatus. This review summarizes some basic principles emerging from these recent studies and proposes a general model for the importance of ribosome pausing in mRNA decay.

## mRNAs have specific half-lives

mRNA decay rates are generally derived by measurement of changes in mRNA labeling kinetics (e.g., pulse chase and approach to steady-state labeling) or by quantitation of the disappearance of mRNA after the inhibition of transcription (Peltz et al., 1991; Peltz and Jacobson, 1993). In the absence of additional transcription the decrease in mRNA concentration vs. time can be expressed as:

$$d[mRNA]/dt = -k[mRNA]$$

where k represents the rate constant for decay. mRNA decay is plotted (on semi-

NATO ASI Series, Vol. H 97
Post-transcriptional Control of Gene Expression
Edited by Orna Resnekov and Alexander von Gabain
© Springer-Verlag Berlin Heidelberg 1996

log axes) as the decrease in relative mRNA abundance vs. time (Fig. 1). The unit of decay is half-life ($t_{1/2}$), i.e., the time required for the disappearance of half of the mRNA. In a given cell, most individual mRNAs decay with characteristic linear (first order) decay kinetics that may differ from each other by as much as 10- to 100-fold (Peltz et al., 1991; Peltz and Jacobson, 1993; see Fig. 1). Such differences in the inherent stabilities of individual mRNAs provide an explanation for the complex decay kinetics of total cellular mRNA.

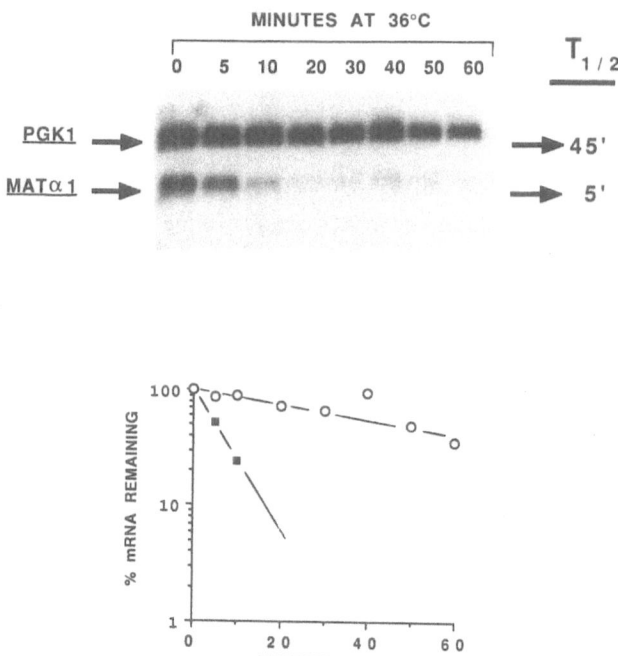

**Figure 1.** Measurement of mRNA decay rates in the yeast *Saccharomyces cerevisiae*. TOP: Transcription is inactivated rapidly by a shift from 24°C to 36°C in cells harboring the *rpb1-1* temperature-sensitive RNA polymerase II mutation. After transcription has been inhibited, decay rates of individual mRNAs are measured by northern analysis of RNA samples isolated at different times after the temperature shift. In the example shown, the blot has been probed for the *PGK1* and *MATα1* mRNAs. BOTTOM: Quantitation of the northern blot. mRNA levels are normalized to the level at t = 0 and half-lives are obtained from the 50% intercept of the lines. Filled squares, *MATα1*, open circles, *PGK1*. From Parker and Jacobson (1990).

**Differences in mRNA decay rates can have significant effects on the expression of specific genes**

Since the instantaneous concentration of any mRNA is a function of the rate constants for both synthesis and decay, differences in the decay rates of individual mRNAs can affect profoundly the overall levels of expression of specific genes. At a minimum, these effects will reflect the large differences in the inherent decay rates of individual mRNAs within a given cell. However, decay rates can also be regulated, for example as a consequence of autogenous feedback mechanisms, the presence of iron or specific hormones, a particular stage of differentiation or the cell-cycle, heat-shock, or viral infection (Atwater et al., 1990; Cleveland, 1988; Peltz et al., 1991). Such changes in decay rates can amplify the effects of parallel changes in transcription rates, leading to very substantial increases (or decreases) in mRNA levels (Hargrove and Schmidt, 1989). In addition to these effects, differences in mRNA decay rates can be important determinants of the time required to achieve changes in the steady-state. Upon a change in transcription rate, the time required to reach a new steady-state will be fastest with an unstable mRNA (Hargrove and Schmidt, 1989). Moreover, while changes of the same magnitude in transcription rates or mRNA decay rates may ultimately lead to the same new steady-state level of mRNA, the time required to reach the new steady-state level can differ dramatically for the two processes (Hargrove and Schmidt, 1989). Clearly, variations in mRNA decay rates offer the cell considerable flexibility in varying the concentrations of mRNAs and, ultimately, the concentrations of proteins.

**mRNA turnover is intimately linked to mRNA translation**

A large set of observations point to an important role for translation in the mRNA decay process. Evidence for this linkage comes from experiments which show that: a) drugs or mutations that interfere with translational elongation promote mRNA stabilization, b) sequence elements that dictate rapid mRNA decay can be localized to mRNA coding regions and the activity of such elements depends on ribosome translocation up to, or near the element, c) premature translational termination can enhance mRNA decay rates, d) degradative factors can be polysome-associated and e) metabolism of the poly(A) tail, a structure recently shown to be involved in translational initiation, is a rate-limiting step in the decay of several mRNAs (Higgins et al., 1992; Peltz et al. 1991; Peltz and Jacobson, 1992, 1993). A requirement for translation may not be limited to a single step in the turnover process, but may reflect a need for ongoing protein synthesis at several mutually dependent events (e.g., a metabolically unstable nuclease <u>and</u> a requirement for ribosome activation of the nuclease--see below).

**cis-acting sequences dictate inherent mRNA decay rates**

As opposed to earlier suggestions that general features such as mRNA size might govern random interactions with non-specific nucleases, current evidence suggests that mRNA turnover is a very specific process requiring cis-acting sequences, trans-acting factors, and, as noted above, ongoing translation. The rapid decay of unstable mRNAs can be explained by the presence within these mRNAs of specific sequence elements which promote their recognition by the cellular turnover machinery. Such "instability elements" have been found within unstable mRNAs of several different eukaryotic organisms, and are located in both coding and non-coding regions (Higgins et al., 1992; Peltz et al., 1991; Peltz and Jacobson, 1992, 1993). Some mRNAs appear to contain more than one element.

The functions of most of these elements are still largely unknown. Data from several labs suggests that one type of element (e.g., the AU-rich element [ARE] found in the 3'-untranslated regions of proto-oncogene, cytokine, and lymphokine factor mRNAs) promotes poly(A) removal which, in turn, destabilizes these mRNAs, possibly by making them suitable substrates for other ribonucleases (Brewer and Ross, 1988; Peltz et al., 1991; Shyu et al., 1991). Other instability elements may be sites of endonucleolytic cleavage, binding sites for trans-acting factors, or sites of ribosome pausing (Bernstein et al., 1992; Peltz et al., 1991; Peltz and Jacobson, 1992, 1993). A paused ribosome or a trans-acting factor bound to a specific site may stimulate mRNA decay indirectly, i.e., by activating a nuclease or promoting an interaction with a nuclease (see below). Evidence for such indirect effects follows from the experiments of Cleveland and colleagues which showed that $\beta$-tubulin mRNA is destabilized by a co-translational protein:protein interaction in which free tubulin subunits bind the nascent tubulin polypeptide as it emerges from the ribosome (Cleveland, 1988).

Further evidence for such indirect effects of destabilizing sequences has been uncovered in studies of nonsense-mediated mRNA decay, the process by which premature nonsense mutations promote rapid decay of otherwise stable mRNAs. Experiments from Peltz et al. (1993) have shown that nonsense-mediated decay of the yeast *PGK1* mRNA requires both a nonsense codon and a downstream destabilizing element. Within the downstream element, two AUGs bracketed by identical nucleotides are crucial for destabilization. These experiments have led us to propose that nonsense-mediated mRNA decay is activated by translational reinitiation per se or by the ribosome pause caused by either the reinitiation event or by a base-pairing interaction between the reinitiation site in the *PGK1* mRNA and a segment of 18S rRNA (Peltz et al., 1993).

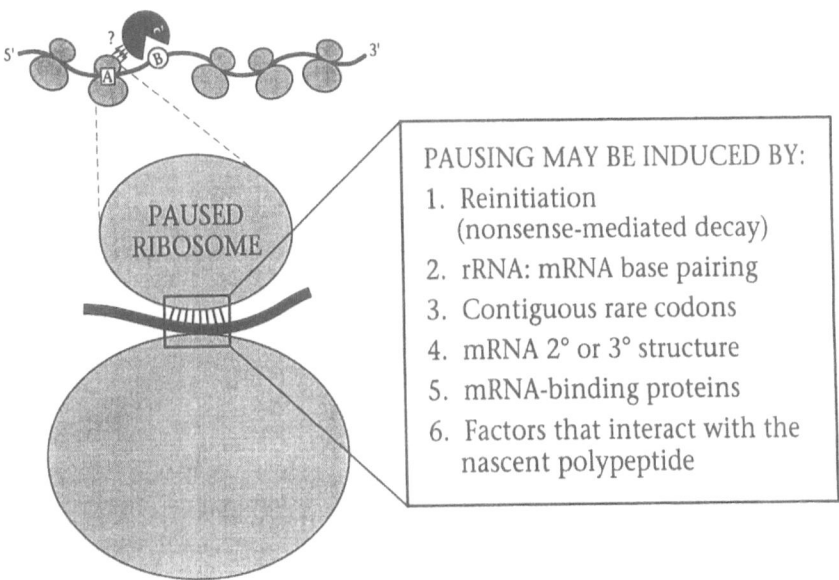

**Figure 2. Model for ribosome-activated mRNA decay.** The cartoon at the top depicts a ribosome paused at site A activating or recruiting an endoribonuclease which then cleaves at site B. The location of site B 3′ to site A, and within the coding region, is not essential. Site B can be located anywhere within the mRNA (and possibly on a different mRNA molecule). Possible reasons for the specific pausing of the ribosome at site A are shown in the enlarged portion of the model.

## A generic role for ribosome pausing in mRNA decay

As noted above, mRNA turnover requires specific sequences and factors as well as ongoing translation. In the simplest models, the factors could be ribonucleases, the specific sequences their targets, and the requirement for translation could reflect a need to replenish the pool of relatively unstable nuclease molecules. However, little evidence has accumulated to support the tenets of this hypothesis. Given the paucity of data to suggest that instability elements are cleavage sites or that the factors in question are actually nucleases, and alternative explanations for the role of the ribosome, we suggest a general model for ribosome-activated mRNA decay. In this model (see Fig. 2) a ribosome must pause at a specific site, i.e., pausing per se is insufficient. Pausing may be a consequence of translational reinitiation or rRNA:mRNA base pairing (as in nonsense-mediated decay; Peltz et al., 1993), clustered rare codons (as seen in yeast and mammalian instability elements; Bernstein et al., 1992; Peltz and Jacobson, 1993), specific mRNA structures or binding proteins (Peltz and Jacobson, 1992), or interactions promoted by the nascent polypeptide (Cleveland, 1988). Within the context of a specific site, we propose that the paused ribosome would activate a nuclease, expose a downstream (or upstream) cleavage site, or promote poly(A) shortening. If cleavage is mediated by an unstable nuclease the requirement for translation would encompass at least two steps in the turnover process.

## Acknowledgments

This work was supported by grants to S.W.P. (GM48631) and A.J. (GM27757) from the National Institutes of Health and by an American Cancer Society Junior Faculty Award (B-67209) to S.W.P.

## Literature Cited

Atwater, J.A., R. Wisdom, and I.M. Verma. 1990. Regulated mRNA stability. Ann. Rev. Genet. 24:519-541.

Bernstein, P.L., D.J. Herrick, R.D. Prokipcak, and J. Ross. 1992. Control of *c-myc* mRNA half-life *in vitro* by a protein capable of binding to a coding region stability determinant. Genes and Devel. 6:642-654.

Brewer, G. and J. Ross. 1988. Poly(A) shortening and degradation of the 3' A + U-rich sequences of human *c-myc* mRNA in a cell-free system. Mol. Cell. Biol. **8**:1697-1708.

Cleveland, D.W. 1988. Autoregulated instability of tubulin mRNAs: a novel eukaryotic regulatory mechanism. TIBS **13**:339-343.

Hargrove, J.L., F.H. Schmidt. 1989. The role of mRNA and protein stability in gene expression. FASEB J. **3**:2360-2370.

Higgins, C., S.W. Peltz, and A. Jacobson, A. 1992. mRNA Turnover in Prokaryotes and Lower Eukaryotes. Curr Opin. Genet. Devel. **2**:739-747.

Parker, R. and A. Jacobson. 1990. Translation and a forty-two nucleotide segment within the coding region of the mRNA encoded by the *MATα1* gene are involved in promoting rapid mRNA decay in yeast. Proc. Natl. Acad. Sci. USA **87**:2780-2784.

Peltz, S.W., G. Brewer, P. Bernstein, and J. Ross. 1991. Regulation of mRNA turnover in eukaryotic cells. Crit. Rev. Euk. Gene Exp. **1**:99-126.

Peltz. S.W., A.H. Brown, and A. Jacobson. 1993. mRNA destabilization triggered by premature translational termination depends on at least three *cis*-acting sequence elements and one *trans*-acting factor. Genes and Devel. (in press).

Peltz, S.W. and A. Jacobson. 1993. mRNA turnover in *Saccharomyces cerevisiae*. In *Control of mRNA stability* (Eds. J. Belasco and G. Brawerman), Academic Press, N.Y. pp. 291-328.

Shyu, A.-B., J.G. Belasco, and M.E. Greenberg. 1991. Two distinct destabilizing elements in the *c-fos* message trigger deadenylation as a first step in rapid mRNA decay. Genes and Devel. **5**:221-231.

# ENZYMES INVOLVED IN CONTROL OF mRNA DECAY IN *E.COLI*

Marianne Grunberg-Manago* and Alexander von Gabain**


**Karolinska Institute, Dept. of Bacteriology, Box 60400, 104 01 Stockholm, Sweden

## A. GENERAL BACKGROUND

The level of expression of a gene depends on the rate of synthesis of its mRNA, the translational efficiency of the mRNA and also the rate of decay of the messenger RNA. While much is known about transcription and translation, the factors that determine the rate of mRNA decay are less well understood. It was believed for a long time that all prokaryotic mRNAs were unstructured and very unstable. We know now, that many bacterial messenger RNAs have regions of complex structure often responsible for the regulation of their expression (e.g. : mRNAs for the ribosomal proteins, phage T4 gene 32 mRNA, aminoacyl-tRNA synthetase mRNAs) and *ompA*). Furthermore, the half-life of the different mRNAs varies 40 fold in *E. coli*; from less than 30 seconds to more than half an hour. This contributes to the differential expression of individual genes. The average life time of mRNA in *E. coli* is so short that the decay of long polycistronic mRNAs may start at their 5' end could start before the 3' end is synthesized. These properties mean that messenger degradation is an alternative method to permit rapid adaptation of bacterial gene expression to changes in growth conditions. RNA turnover is a major metabolic activity of the cell. 50-60% of the total RNA synthesis is devoted to mRNA but because of its rapid degradation mRNA constitutes only 2-3% of the bulk cellular RNA. For reasons of economy this energy consuming turnover should be carefully regulated. A number of observations suggest that structural features of mRNAs determine their stability. It is important to distinguish between the functional and chemical stability of mRNAs. The chemical stability reflects the average time that the mRNA is detected by some kind of hybridisation measurement. The functional stability is a measure of the average time that a given species of mRNA is active directing the synthesis of a func-

*Institute de Biologic Physico-Chimique, 13 Pierre et Marie Curie, 7500 Paris, France and **Vienna Biocenter, Dept. ot Microbiology, Dr. Bohrgasse 9, A-1030 Vienna, Austria.

NATO ASI Series, Vol. H 97
Post-transcriptional Control of Gene Expression
Edited by Orna Resnekov and Alexander von Gabain
© Springer-Verlag Berlin Heidelberg 1996

tional and structurally complete polypeptide. Although about 20 RNases have been identified in *E. coli* (Deutscher, 1985) only a few have been shown to be involved in mRNA stability. Most of the bacterial RNases that have been characterised are implicated in the maturation of long RNA precursors into tRNA and rRNA. No single model for mRNA inactivation has emerged; several different mechanisms acting independently on different target sites may inactivate a given mRNA. The decay of mRNA proceeds through the action of both endonucleases and exonucleases. Exonucleases attack the free ends of an RNA molecule while endonucleases cleave the RNA internally. Only exonucleases which act processively 3'-> 5' are known in *E. coli*. This implies that early degradation at the 5' end of a mRNA must be initiated by an endonuclease followed by exonuclease digestion at the 3' end at the endonucleolytic cut.

## B. EXONUCLEOLYTIC ATTACK

**1. PNPase and RNase II:** In *E. coli* two exonucleases are known to participate in the 3'->5' decay of mRNA : polynucleotide phosphorylase "PNPase" and RNase II (see Table 1). These two enzymes can substitute for each other in the degradation of RNA; mutants in either RNase II or PNPase are viable, the double mutants however are not viable (Donavan and Kushner, 1986). In *E. coli* crude extracts there is about 10 times more RNase II than PNPase activity (C. Portier, personal communication). Both enzymes attack from the 3'OH extremity to the 5' end. So far no exonucleases able to degrade from the 5' end have been identified in bacteria. PNPase is a phosphorolytic enzyme; the product is a nucleotide diphosphate and the reaction is reversible. Since PNPase has a low affinity for diphosphates and a high affinity for polymers, its role in vivo is probably degradative and not synthetic. PNPase is a trimer of 77KDa subunits, however a more complex form can be isolated as a pentamer corresponding to 320-360 Kda. (C. Portier, 1975). RNase II is a hydrolytic enzyme; the end products are monophosphates. PNPase is widely distributed in bacteria but has not been found in other organisms. RNase II is less widely distributed; it has not been found in *B. subtilis* (Deutscher and Reuven, 1991). RNase II is a monomer of 64KDa by sequence (Zilhao et al., 1993). Both enzymes degrade long RNA polymers processively (Godefroy et al., 1972, Nossal and Singer, 1968). In contrast, oligonucleotides are degraded in a synchronous manner by PNPase and RNase II. It was suggested as a possible explanation of this result that PNPase has two binding sites for long polymers during phosphorolysis (Godefroy, 1970 Fig. 1).

**TABLE 1**   *E. coli* RNases implicated in mRNA degradation

---

| RNase | map | size | reference |
|-------|-----|------|-----------|

---

*Exonucleases*

| RNase | map | size | reference |
|-------|-----|------|-----------|
| PNPase | 69' | trimer 77.122 kDA pentamer 320-360 kDA | Godefroy-Colburn and Grunberg-Manago, 1972 |
| RNase II | 28' | 64kDa | Gupta et al., 1977 Nossel and Singer, 1968 Zilhao et al., 1993 |

*Endonucleases*

| RNase | map | size | reference |
|-------|-----|------|-----------|
| RNase III | 55' | dimer of 25 kDa | Robertson et al., 1968 Dunn, 1976 |
| RNase I | 14' | 27 kDa | Spahr and Hollingsworth, 1968 Neu and Heppel, 1964 |
| RNase I* | | 27 kDa | Cannistraro and Kennel, 1991 |
| RNase M | | 27 kDa | Cannistraro and Kennel, 1989 |
| RNase IV | | 31 kDa | Spahr and Gesteland, 1968 |
| RNase F | | ? | Gurevitz et al., 1982 |
| RNase N | | 120 kDa | Misra and Apirion, 1978 |
| RNase E (ams) | 23,5' | 114 kDa | Apirion, 1978 Babitzke and Kushner, 1991 Casaregolo et al., 1992 |
| RNase K | | 50-60 kDa | Lundberg et al., 1990 |
| Reg B | T4 enzyme | | Uzan, Favre and Brody, 1988 Sanson and Uzan, 1993 |

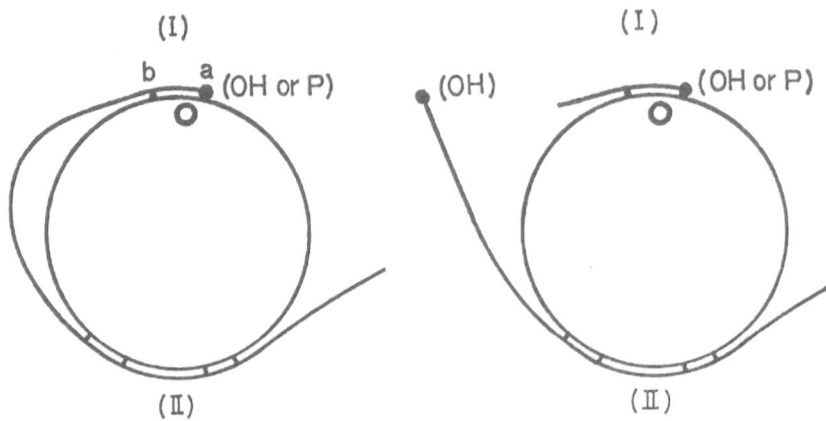

Fig.1
Possible schemes of polymer binding to polynucleotide phosphorylase during phosphorolysis.(A) Phosphorolysis of a long polymer, or inhibition by a long 3'P-polymer. (B) Simultaneous phosphorolysis of a polymer and an oligonucleotide, or, inhibition of polymer phosphorolysis by a 3'P-oligonucleotide. The black dots represent the 3' end of the polymers. (I) and (II) represent the subsites I and II, and the open circles represent the inorganic phosphate, which is the other substrate of the reaction. From Godefroy, 1970.

The phosphorolysis reaction was proposed to occur in the first region where the 3' OH end of the polymer is bound with a short life time while the bulk of the polymer binds to a second region further away from the active site with a long residence time. The second region was believed to be responsible for the processive degradation but to play no part in the phosphorolysis of oligonucleotides. Both PNPase and RNase II enzymes are inhibited by secondary structure; however once polynucleotide phosphorylase has bound to the free 3'OH end it is able to degrade polymers with as much secondary structure as tRNA (Thang et al., 1967, Beltchev et al., 1971), on the other hand RNase II is stopped by stem-loop structures inside the polymers. There are similar differences in the sensitivity of mRNA to both enzymes *in vivo* and *in vitro* (Guarneros and Portier, 1990). Most mRNA can form some secondary structure at their 3' end (e.g. from Rho independent terminators) and are therefore resistant to the exonucleolytic attack by PNPase and RNase II. Like the hairpin structure found in the Rho independent terminators, secondary structures such as the REP sequences found at the end of the transcripts or between the message of polycistronic mRNA can also protect the upstream message from the attack of 3'->5' exonucleases (Higgins et al., 1988). Processive degradation from the

3' end probably contributes very little, as a initiating event, in the inactivation of most messages. However, exonucleases are necessary to function in conjunction with endonucleases which remove the hairpins and thus destabilise upstream RNA by allowing an entry for exonucleases (Regnier, Personal communication). The recycling of RNA precursors via the degradation of RNA fragments by exonucleases may be crucially important for the maintenance of adequate pools of nucleotide precursors for RNA synthesis.

**2. Expression of exonuclease genes:** The genes *pnp* and *rnb* which are located at 69' and 28' on the *E. coli* chromosome, code for PNPase and RNaseII respectively. The genes have been sequenced and they do not show any evident homology. However the PNPase primary structure shows an homology with the RNA binding domain of ribosomal protein S1 (Regnier et al,, 1987). The regulation of the expression of the gene for RNase II is under study. The regulation of the expression of PNPase is quite complex. The *pnp* mRNA is synthesized from two promoters P1 and P2; the transcripts are processed by RNase III at a site situated between P2 and the *pnp* ribosomal binding site.

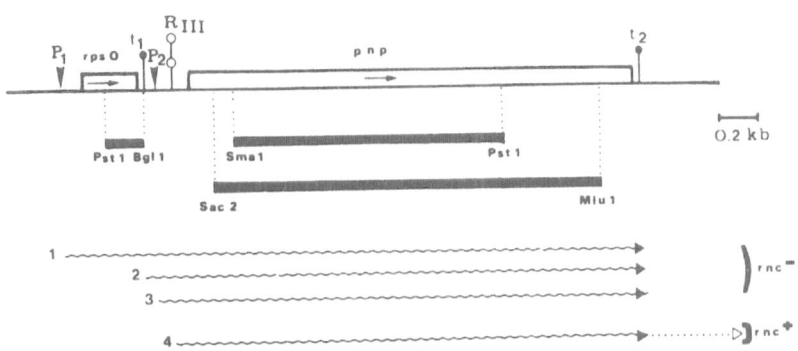

Fig.2
Structural organization of *rpsO-pnp*. Coding sequences, transcription signals and maturation sites in both operons are shown. $P_1P_2$, promoters ; $t_1$ $t_2$, terminators, RIII, RNase III-sensitive site ; horizontal arrows give the direction of transcription; open boxes represent the structural part of the genes. Black boxes beneath the map correspond to the probes used in hybridization experiments; cleavage sites at each extremity are specified. Transcripts previously identified by S1 mapping in the *rpsO-pnp* operon are represented by wavy lines. From Portier et al. 1987.

The transcript from P1 promoter also expresses the *rpsO* product, the gene for ribosomal protein S15 which is upstream of *pnp*. It has been shown with the help of a translational fusion between *pnp* and *lacZ* that, in the presence of increased cellular concentrations of PNPase, the level of the hybrid protein, Pnp-β-galactosidase, is repressed whereas the synthesis rate of the corresponding message is not significantly affected (Robert-Le Meur and Portier, 1992). PNPase expression is therefore post-transcriptionally, negatively auto-controlled. However, this auto-regulation is totally abolished in strains where the RNase III site on the *pnp* message has been deleted or in strains devoid of RNase III. The cleavage by RNase III near the 5' end of the *pnp* message could permit the auto-control by allowing the PNPase enzyme to bind to a end generated by RNase III cleavage. Mutational analysis supports the hypothesis that PNPase, a 3'-5' exonuclease, recognizes a specific repressor binding site at the 5' end of its mRNA. At first this result, that a 3'-5' non-specific exonuclease binds specifically to its own messenger and in particular to the 5' part of it, was surprising. However, it is not in contradiction with a model where there are two different nucleotide binding subsites on the protein of which one binds the 5' end of the polymer (see above). This could imply that the enzyme might have other roles in mRNA stability in addition to its 3'-5' exonucleolytic function.

## C. ENDONUCLEOLYTIC ATTACK

There is a strong experimental evidence that the stability of many mRNA depends on the structure of the 5' end of mRNAs. The nature of the putative wave of processive decay in the 5'-> 3' direction is completely unknown. The phosphorylation state of the 5' end of the mRNA may be an important determinant of the susceptibility to this type of decay : mRNA with a triphosphate were more stable than those with a monophosphate (Lin Chao and Cohen, 1991). Nevertheless some processed transcripts are quite stable, so additional factors such as the primary structure or the secondary structure at the 5' end of the transcript may also influence the rate at which the wave of 5'->3' decay is initiated and this will be discussed below. Few RNases have been implicated in endonucleolytic attack of mRNAs. In some cases different reported activities may actually be due to a single enzyme. The endonucleases which are the most clearly involved in mRNA metabolism are RNase III and the *ams/rne*-controlled nucleases RNase E and RNase K.

## D. RNase III

**1. Cleavage of mRNA:** RNase III was first detected in *E. coli* as an endoribonuclease that cleaves double stranded RNA molecules (Robertson et al., 1968). It was discovered that the enzyme is involved in the processing of stable RNA species, it was also shown that it processes some phage and cellular mRNA molecules (Dunn and Studler, 1973). However, only a subset of the bacterial mRNA population is affected by RNase III processing and the enzyme is not essential; mutants with defective RNase III are viable. RNase III is a dimer of 25 KDa subunits. *In vivo* the enzyme cleaves specific regions of a single RNA that form base paired stem loop structures (Krinke and Wulff, 1987). It also degrades natural duplex RNA molecules that arise by the annealing the antisense and sense RNA molecules (Krinke and Wulff, 1987). The specific structural and sequence elements that define the cutting site remain largely unpredictable. RNA processed by RNAse III are cleaved in one strand or both strands of RNA helix. Whether the product of RNase III action is a single strand nick or a double strand break depends in part upon the degree of base pairing in the region of the cut sites. Krinke and Wulff (1990) used a natural double strand RNA from phage lambda of *E. coli* to examine duplex RNA cleavage in closer detail. RNase III appears to measure and cleave at 10 to 14 base pairs from the edge of the double strand region. By comparing the sequence of more than 20 RNase III processing sites Krinke and Wulff (1990) derived a consensus sequence - however, there remains a great deal of uncertainty.

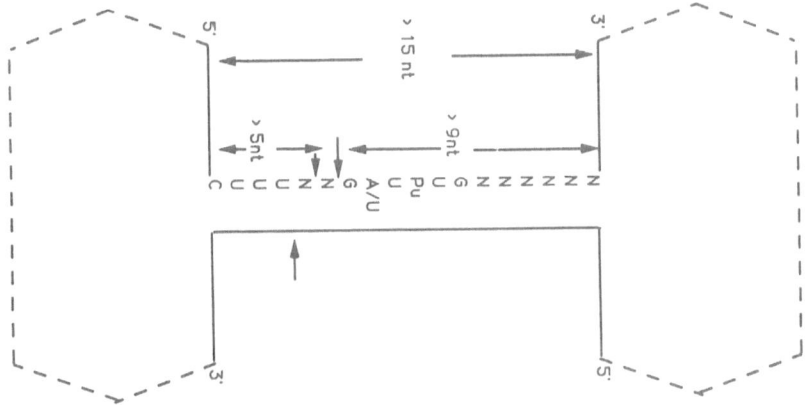

Fig.3
RNase III Processing sites. From Krinke and Wulff, 1990.

The gene *rnc* for RNase III was mapped at 55 min. (Apirion and Watson, 1975) and the sequence of the wild type *rnc* gene has been determined. The *rnc* gene is the first of three genes in an operon (Takiff et al., 1989, Fig. 4) with the *era* and *recO* genes located downstream. The product of the *era* gene is a GTP binding protein with GTPase activity essential for the growth of *E. coli* although its function is still unknown. The product of the *recO* gene takes part in recombination and DNA repair pathway.

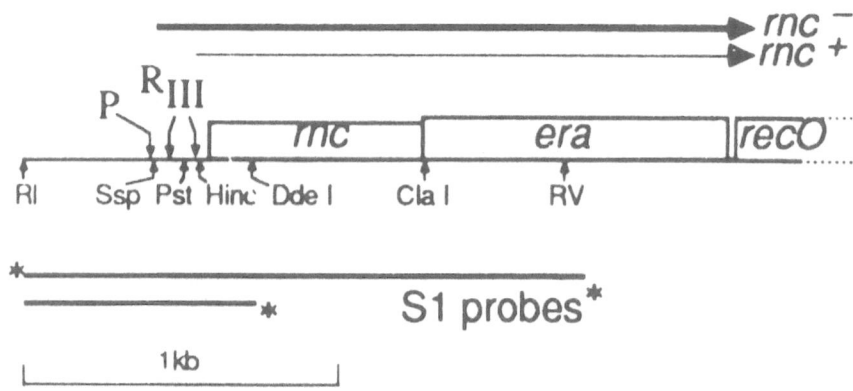

Fig. 4

Transcription and maturation of the *rnc-era* operon. The results of S1 mapping experiments are summarized over the map of the *E. coli* chromosome showing the *rnc, era* and *recO* genes. Transcripts identified by S1 mapping are shown by horizontal arrows of a width proportional to abundance. DNA probes are represented by solid bars. Labeling of 5' ends is shown by a star. The promoter (P) and RNase III (RIII) maturation sites deduced from the S1 nuclease experiments are indicated on the genetic map. Restriction sites are RI, EcoRI; Ssp, SspI; Pst, PstI; HincII; RV, EcoRV. From Bardwell et al., EMBO Journal, 1989.

**2. Regulation of *rnc* expression and RNase III activity:** Several mutations in the *rnc* gene are known, both null mutations caused by mini Tn10 insertions as well as point mutations within *rnc* the structural gene. The absence of RNase III in a cell has pleiotropic effects on the expression of many proteins (Gitelman and Apirion, 1980). This suggests that RNase III has a potential global regulatory effect. In the leader region of the *rnc* operon there is a hairpin structure and it was demonstrated that RNase III processed this site *in vivo* and *in vitro* (Bardwell et al.,1989). In the absence of this processing (*rnc* mutation) there is an increase in the level of the non-functional RNase III and Era proteins. This correlates with an increase in the half-life of the mRNA which is 45-50 sec in the wild type and 5 min in the *rnc* mutant strain. Thus, RNase III controls its own expression and also that of the downstream gene, *era*, in the operon. At least three cases are known where RNase III cleavage is implicated in the rate limiting step that leads to mRNA decay. In addition to the two *E. coli* operons *rnc*, *era*, *recO* and *rpsO*, *pnp*, just described, the *metY*, *nusA*, *infB* operon also contains an RNase III site controlling mRNA stability (Regnier and Grunberg-Manago, 1990, Fig. 5).

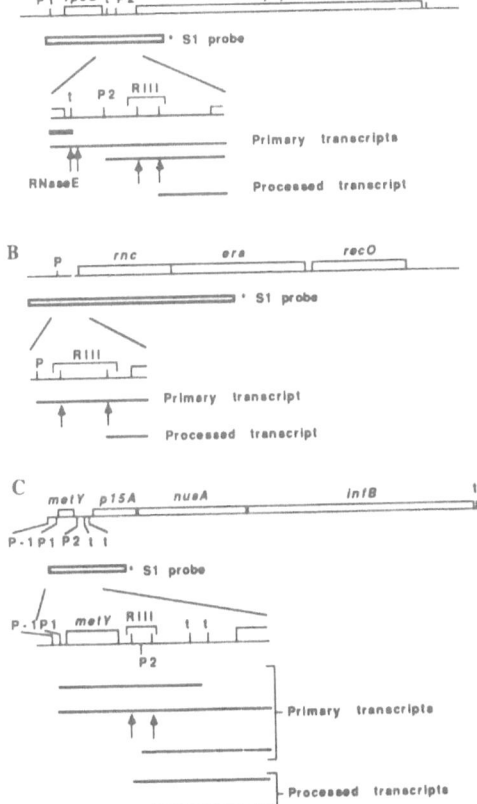

Fig. 5
Characterization of primary and RNase III processed transcripts of the *rpsO-pnp*, *rnc-era-recO* and *metY-nusA-infB* operons. The DNA used for probing the 5'ends of the RNAs are shown by open boxes under the genetic maps of the *rpsO-pnp* (A), *rnc-era-recO* (B) and *metY-nusA-infB* (C) regions of the chromosome indicating the locations of the coding sequences of the genes and the transcription promoters (P-1, P, P1 and P2) and terminators (t). The stars indicate the radioactive end of the DNA strand complementary to the RNA. The structures of the RNA are shown by lines under a magnification of the region of interest. Locations of the RNase III (RIII on the maps) and RNase E processing sites are shown by arrows. From Regnier and Grunberg-Manago, 1990.

This last operon, which is located immediately upstream of *rpsO*, *pnp*, codes for a minor form of initiator tRNA, NusA, a protein involved in termination and IF2, translational initiation factor 2. Within the operon there are genes for three other proteins of unknown function called p15A, p15B and p35. In each of these three operons the cleavage of the primary transcript by RNase III removes RNA secondary structures that can form in a non-coding segment located 5′ of one of the genes.

This processing by RNase III dramatically reduces the half-life of mRNA for the downstream translational units. Removal of the base-paired RNA structure may facilitate the entry of another nuclease, or create a site responsible for an auto-regulation (as in the case of *pnp*). In the IF2 operon, there are two promoters P1 and P2; P2 is located between *metY* and the gene for p15A protein. The RNase III site overlaps the P2 promoter. Transcripts from P2 do not include the stem structure contained in P1 transcripts (Fig. 6).

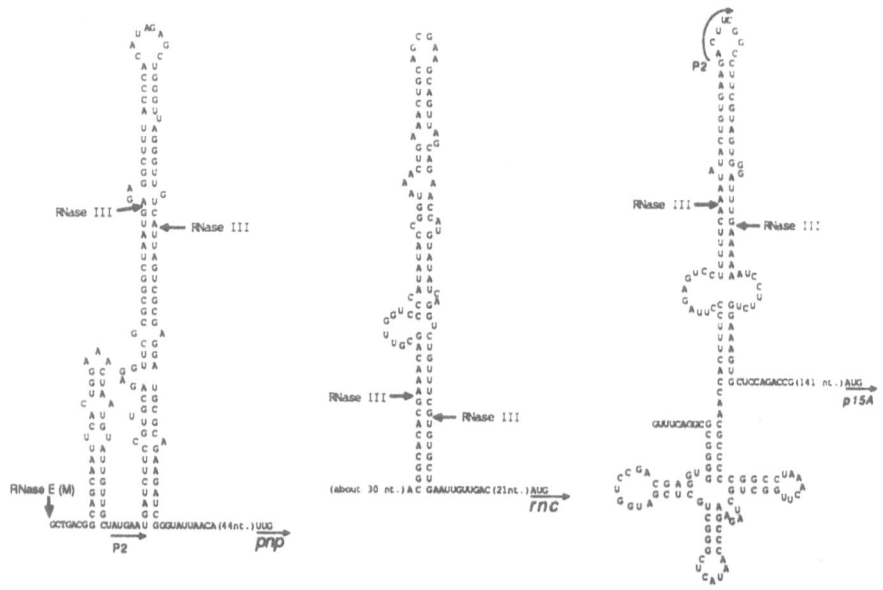

Fig. 6
Presumptive structures in the regions of the 5′RNA leader processed by RNase III. The RNase E maturation site (also called M in the text) in the *rpsO-pnp* co-transcript, the RNase III cutting sites, the secondary promoters of the *rpsO-pnp* and *metY-nusA-infB* operons (P2), and the translation initiation codons are indicated by arrows. From Regnier and Grunberg-Manago M, 1990.

P1 transcripts are sevenfold more stable than P2 transcripts. The P2 transcripts are unprocessed and their stability is unaffected by the presence of RNase III. By contrast, the P1 transcripts are processed by RNase III to give two new transcripts whose 5' ends correspond to the two RNase III cleavage sites on either side of the stem-loop structure. Elimination of the stable stem structure decreases the stability of the two processed transcripts. The 5' end of P2 transcript is located between the two ends of the processed transcript. Both processed RNAs are 20-30 fold less stable than the primary P1 transcript, but they are also less stable than the unprocessed P2 transcript even though the 5' of the P2 RNA is located between those of the two processed P1 mRNAs. This could be because the primary transcripts have a triphosphate at their 5' ends while the processed RNAs have a 5' monophosphate. Another possibility is that RNase III in the cell exists as a complex with other nucleases so that RNase III cleavage may trigger the attack by other nucleases at the processed ends. The decreased stability of the RNase III processed mRNA does not result in a decreased expression of NusA or IF2. On the other hand, in both the *pnp* and *mc* operon there is a decreased expression of functional proteins located downstream of the cleavage site. The possibility exists that in addition to triggering the decay of specific RNAs, RNase III affects translation in more direct ways (as shown in the *pnp* mRNA) by changing the structure of mRNA and affecting its ability to initiate translation.

**3. Activation of lambda N gene and T7 genes expression:** RNase III cleavage could activate translation if a ribosome binding site hidden in the messenger becomes accessible to ribosomes after cleavage. This occurs in the case of the translation of the N gene of phage l as well as the 0.3 gene of phage T7 (Dunn and Studier, 1975). The N gene transcript contains a long leader RNA upstream of the structural gene which forms a secondary structure sensitive to endonucleolytic processing by RNase III. RNase III processing of the N leader in wild type cells enhances N expression three fold compared to N expression in *mc* strain. The processing has little effect on N mRNA stability, but does make the ribosome binding site available and so enhances translation (Court, 1993). The leader RNA of 0.3 gene of T7 is 175 nucleotides long and like the N leader contains a stem structure sensitive to RNase III. The mRNA levels are not affected by processing. The processed 0.3 transcript is more efficiently translated in vitro and in vivo (Court, 1993). The different effect of RNase III processing, either inhibition or activation depends on the stability of mRNA and the accessibility of the ribosome binding site.

**4. Inhibition of gene expression by RNase III processing within the 3' non coding region:** A good system to study the effect of a combination of the exonucleases, PNPase and RNase II, with the endonuclease RNase III on mRNA degradation is the expression of *int* gene of bacteriophage lambda. At different stages of the lambda life cycle, the *int* gene is transcribed from two different promoters PL and PI but only the PI transcript is translated (Guarneros, 1988). When transcription starts at the PI promoter *int* mRNA ends at t1, a terminator 276 nucleotides beyond the stop codon of the *int* gene; this mRNA expresses Int protein efficiently.

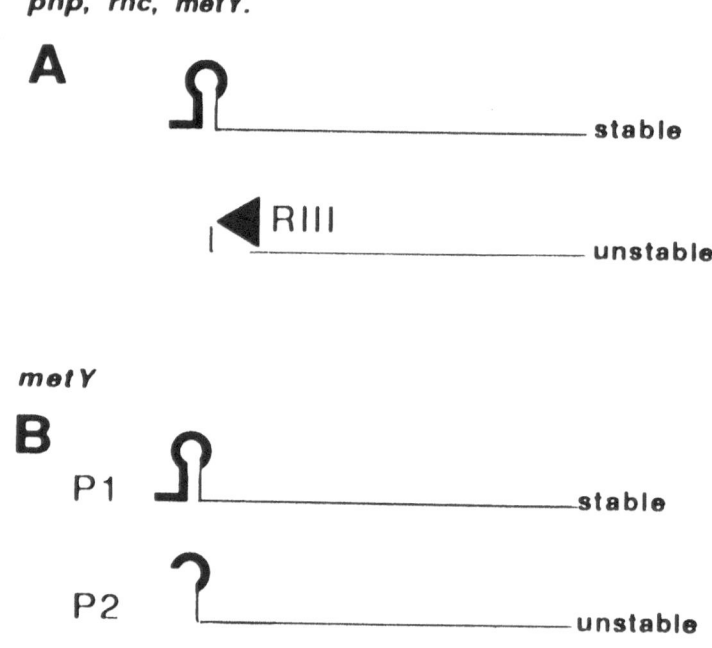

**Fig. 7**
Model of stabilization of mRNAs by a 5' structural motif. Part (A) of the figure represents the structure of the stable primary transcripts of the *rpsO-pnp, rnc-era-recO* and *metY-nusA-infB* operons containing the RNase III hairpin compared to that of the unstable RNase III processed transcripts missing this feature. Part (B) shows the structures of the stable P1 and unstable P2 primary transcripts of the *metY-nusA-infB* operon. We propose that 5' terminal motifs absent in RNase III processed and in the P2 initiated RNAs are responsible for the stability of the primary transcripts. These motifs, represented by the thick part of the lines, require the presence of the RNase III hairpin and/or of the upstream RNA. From Regnier and Grunberg-Manago, 1990.

The stem of the terminator is important for *int* mRNA stability: the lower the mRNA stability, the lower is the expression of *int* gene. Mutations which inactivate PNPase (but not RNase II) appreciably increase the stability of the *int* mRNA. Thus the stability of the PI mRNA is determined by the secondary structure at its 3′ end and the presence of an active PNPase.

On the other hand transcription of *int* from the other lambda promoter PL is accompanied by transcription antitermination by the N protein coded in the transcript from PL. This transcript traverses the t1 terminator and forms a different elongated stem-loop structure which has been named Sib. This structure is sensitive to RNase III cleavage. The initial cleavage by RNase III is followed by rapid 3′-5′ exonuclease processing of the upstream mRNA which prevents its translation into active protein. This control of gene expression has been named negative "Retro regulation". Cells that are defective in RNase III do not process sib and therefore express *int* from pL. Cells deficient in PNPase but not RNase II also express int from pL promoter (Guarneros and Portier, 1990).

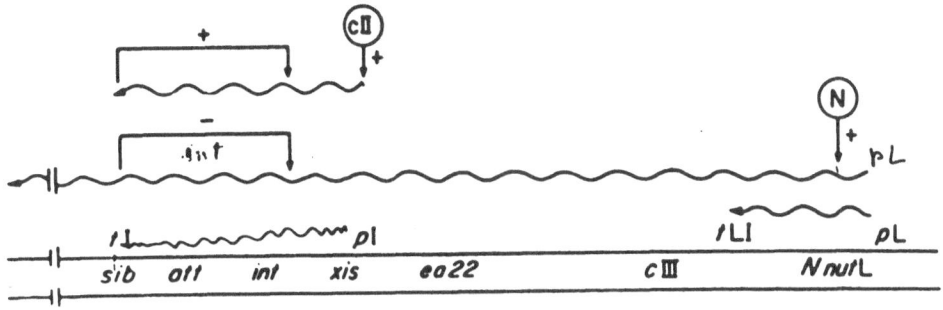

Fig. 8

Regulation of int gene expression. Immediately after infection, transcription initiated at pL terminates at tL1. Once active N protein is available, it acts at the site nutL and transcription initiated at pL proceeds through tL1 and other terminators. The int gene is transcribed but is not expressed from the pL transcript because the regulator sib prevents int mRNA translation into an active Int product. Promoter p I is activated by cII protein later in infection. The pI transcript terminates at tI and is translated efficiently into Int protein. The parallel lines represent a segment of I DNA ; several genetic markers have been positioned between the lines. t I overlaps *sib*, and pI partly overlaps xis in the DNA. The wavy arrows indicate origin, direction, and extent of I transcripts. The straight arrows, flanked by+or- signs, indicate stimulatory or inhibitory activities. Encircled N or cII symbolize the respective I-encoded proteins. From Guarneros and Portier, 1990.

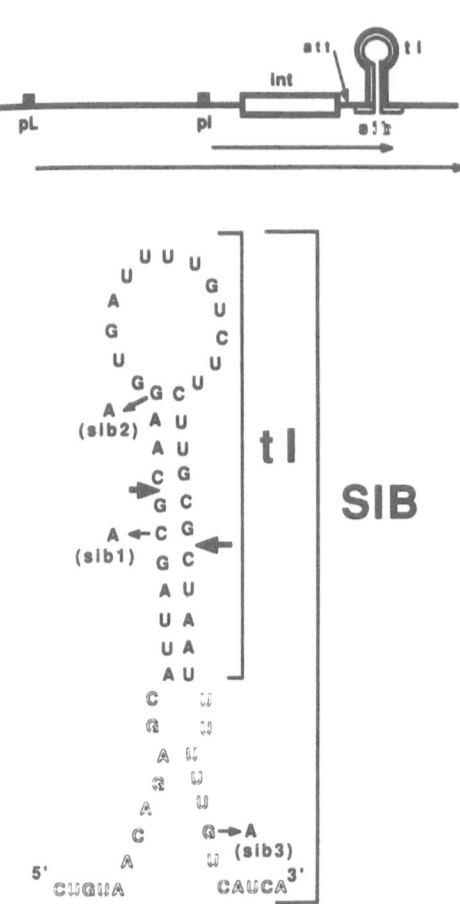

Fig. 9 Top: the transcription modes of the lambda *int* gene : the open bar represents the *int* structural gene. The positions of the pL and pI promoters and the attachment site (att) are indicated. The horizontal arrows indicate the direction and extent of the transcripts . The transcript from *pI* stops at *t* I, the pL transcript continues through t I creating the sib structure. Bottom: overlap of the t I terminator and the *sib* sites. The Sib structure is cleaved by RNase III (large arrow-heads). The position and base substitutions of the sib mutations which abolish RNase III sensitivity are also indicated. From Guarneros and Portier, 1990.

## E. RNase E AND THE *ams/rne* CONTROLLED CLEAVAGES

**1. The *ams^ts* and *rne^ts* mutations:** In order to find mutations that impair mRNA degradation, Kuwano *et al.* (1977) mutagenised *E. coli*, scored for conditional mutants temperature-sensitive for growth, and subsequently identified mutant strains with retarded decay of bulk mRNA at 42°C. The mutation (*ams^ts*) was attributed to a locus called *ams* (*altered mRNA stability*). The phenotype was not an indirect manifestation of impaired translation, as protein synthesis was maintained even at the nonpermissive temperature (Ono and Kuwano, 1979).

The temperature-sensitive *rne^ts* mutation was originally defined by the loss of capacity to process 9S rRNA into 5S RNA at 42°C (Ghora and Apirion, 1978). It was suggested that the mutation resides in a gene encoding an endoribonuclease, designated RNase E. RNase E cleavage has subsequently been demonstrated *in vitro* for various RNAs, including 9S rRNA and RNA I, the antisense RNA that regulates col E1 plasmid replication (Table 2). The cleavage activity of RNase E has been found to be thermosensitive in extracts prepared from *me*ts mutants (Misra and Apirion, 1980; Tomcsanyi and Apirion, 1985). *In vivo* studies with an *me*ts strain subsequently revealed that the decay and the processing of certain mRNA species is impeded at the nonpermissive temperature. This finding that RNase E can control mRNA degradation suggested a more general role for this enzyme (Mudd *et al.*, 1988; Carpousis *et al.*, 1989; Mudd *et al.*, 1990a; Regnier and Hajnsdorf, 1991).

**2. The *ams/rne* gene:** As both the *ams*ts and *me*ts mutations inhibit mRNA degrada-tion, it was of interest to see how they are related. Indeed comparative studies showed that they have almost identical phenotypes, i.e. longer mRNA half-lives and inhibited 9S RNA processing (Mudd *et al.*, 1990b; Melefors and von Gabain, 1991). In four studies (Mudd et al., 1990b; Babitzke and Kushner, 1991; Taraseviciene et al., 1991, Melefors and von Gabain, 1991) it was concluded that *ams*ts and *me*ts are mutations in the same gene, referred to hereafter as the *ams/rne* gene. A recent sequence analysis of the two *ams/rne* alleles confirmed in fact that *ams*ts and *me*ts are point mutations clustered within a stretch of three codons of the putative *ams/rne* gene product (McDowall et al., in press, Casaregola et al., 1992).

The first attempt to clone the originally designated *ams* (i. e *rne*) gene by com-plementation led to the isolation of a "false" clone that turned out to be a part of the *groEL* gene, which encodes a chaperonin protein (Chanda *et al.*, 1985). A DNA fragment from *E. coli* was cloned on the basis of its ability to complement the *me*ts mutation (Dallmann *et al.*, 1987) which we can identify today as a fragment of the *ams/rne* gene. Some years later, a clone comprising the *ams/rne* gene was reported by Claverie-Martin *et al.* (1989). The characterization of the *ams/rne* gene and its gene product took a new turn when a study was initiated with the aim to identify a myosin-like contractile protein in *E.coli* (Casaregola et al., 1990). When us-ing a monoclonal antibody against the heavy chain myosin from Saccaromyces cerevisiae, a protein, Hmp1 (high molecular protein) was identified that migrates as a 180 kDa band on a denaturing SDS gel (Casaregola et al., 1990). Cloning and sequencing the *hmp1* gene encoding a 114 kDa protein, revealed that it is identical with the previous isolated *ams/rne* gene (Casaregola et al., 1992). However, compared with the previous published *ams/rne* sequences (Claverie-Martin *et al.*, 1991; Chauhan *et al.*, 1991), *hmp 1* specifies an ORF which is longer at its C-terminal region (Casaregola et al., 1992). In the study it also has been demonstrated that the

**Table 2** Nucleotide sequence of *ams/rne*-dependent cleavage sites. Left column depicts the designation of the *ams/rne*-dependent cleavage sites. Superscripts indicate reference articles. In cases where the ribonuclease that cleaves at the site *in vitro* has been identified, the identity of the ribonuclease is stated in parentheses. The sequence gap shows the reported site of cleavage. Right columns indicate whether the resulting upstream and downstream cleavage products are stable or unstable *in vivo* relative to the RNA precursor. (nd); not determined.

1, Lundberg, 1991; 2, Lundberg et al., 1990; 3, Roy & Apirion, 1983; 4, Tomcsanyi & Apirion, 1985; 5, Ehretsmann et al., 1992; 6, Pragai & Apirion, 1983; 7, Srivastava et al., 1990; 8, Gurewitz et al., 1983; 9, Carpousis et al., 1989; 10, Brun et al., 1990; 11, Faubladier et al., 1990; 12, Kokoska et al., 1990; 13, Regnier & Hajnsdorf, 1991; 14, Mackie, 1991; 15, Nilsson & Uhlin, 1991.; 16, Gross, 1991; 17, Baumeister et al., 1991; Lin Chao and Cohen, 1991.

| SUBSTRATE | CLEAVAGE SITE | CLEAVAGE PRODUCT | |
|---|---|---|---|
| | | UPSTREAM | DOWNSTREAM |
| *ompA* A[1] (RNase K) | UCAGA CUUUA | UNSTABLE | UNSTABLE |
| *ompA* C[2] (RNase K) | GAAGG AUUUA | UNSTABLE | UNSTABLE |
| *ompA* D[2] (RNase K) | GGCGU AUUUU | UNSTABLE | UNSTABLE |
| 9Sa[3] (RNase E) | ACAGA AUUUG | UNSTABLE | STABLE |
| 9Sb[3] (RNase E) | AUCAA AUAAA | STABLE | UNSTABLE |
| RNAI (RNase E) | ACAGU AUUUG[4] | UNSTABLE[18] | UNSTABLE[18] |
| T4GENE32[5] (RNase E) | UGCGA AUUAU | UNSTABLE | STABLE |
| T4D24[6] (RNase F) | nd | nd | nd |
| 10Sa[7] (RNase III) | CAGCU CCACC | STABLE | UNSTABLE |
| 10Sb (M1)[8] | CACCU GAUUU | STABLE | UNSTABLE |
| T4 GENE59[9] | CUAUG AUUAA | UNSTABLE | STABLE |
| *gltx*[10] | CCAGG AUUUG | UNSTABLE | STABLE |
| *gltx*[10] | CUUAA UUUUU | UNSTABLE | STABLE |
| *dicF*[11] | UCAAU UUUCU | UNSTABLE | STABLE |
| Phage f1 C[12] | AAAAC UUCUU | STABLE | STABLE |
| Phage f1 D[12] | UUAUG UAUCU | STABLE | STABLE |
| Phage f1 E[12] | UGAAU CUUUC | STABLE | STABLE |
| Phage f1 F[12] | UAGAU UUUUC | STABLE | STABLE |
| S15 M[13] | UUCAA GCUGA | UNSTABLE | UNSTABLE |
| S15 M2[13] | GCGAG UUUCA | UNSTABLE | UNSTABLE |
| S20 (191)[14] | UAAGC ACAAC | UNSTABLE | UNSTABLE |
| S20 (301)[14] | AACCG AUCGU | UNSTABLE | UNSTABLE |
| *papB*-A[15] | UUUGU AUUGA | UNSTABLE | STABLE |
| *atpE* 1[16] | UUAAU UUACC | UNSTABLE | UNSTABLE |
| *atpE* 2[16] | UACGU UUUAA | UNSTABLE | UNSTABLE |
| *tetR*[17] | nd | nd | nd |

"180 kDa" polypeptide is the primary gene product and not due to aggregation with other molecules. The sequence data indicate that the highly charged C-terminal region is responsible for its anomalous behavior when analyzed by SDS/PAGE (Casaregola et al., 1992). It seems therefore likely that the *hmp 1* gene is the complete *ams/rne* gene.

The *ams/rne* gene specifies a transcript of about 3.5 kb, as determined by Northern blot analysis (Melefors, 1991). In contrast to *ompA* mRNA, this transcript can barely be detected in wild-type cells or in *ams/rne* mutant strains at the permissive temperature, but it accumulates to relatively high levels in the mutant strains at the nonpermissive temperature (Melefors, 1991; Harde et al., submitted). The promoter of the *ams/rne* gene has been identified; the site of transcription initiation maps 0.36 kb upstream of the *ams/rne* translation initiation codon (Claverie-Martin *et al.*, 1991). It has been suggested that the low level of expression of this gene may, at least in part, be a consequence of post-transcriptional feedback regulation involving cleavage by RNase E (Claverie-Martin *et al.*, 1991). In addition, the *ams/rne* transcript contains a poor Shine-Dalgarno sequence that may be occluded by secondary structure (Claverie-Martin *et al.*, 1991; Chauhan *et al.*, 1991).

**3. Functions of the *ams/rne* gene:** In respect with the function of the *ams/rne* gene product it is worthwhile mentioning that the $rne^{ts}$ mutation has been found to impair cell division (Goldblum and Apirion, 1981) and that in a new report an extensive sequence similarity was discovered between the *ams/rne* gene and the *mre* gene which is implicated in determining bacterial cell shape (Casaregola et al., 1992; McDowall et al., in press). Sequence homologies with myosins support the notion that the *ams/rne* gene product may take part in contractile functions (Mc Dowall et al., in press). Analysis of the sequence of the *ams/rne(hmp1)* ORF disclosed a number of interesting features that suggest multiple function for the Ams/Hmp1 protein (Claverie-Martin *et al.*, 1991, Casaregola *et al.*, 1992). A putative nucleotide binding site and a transmembrane domain was identified in the N-terminal half of the Ams/Hmp1 molecule (Casaregola *et al.*, 1992). In the carboxy-terminal amino acids a homology of 18% was found between Ams/Hmp1 and a ribosomal protein from *Neurospora crassa* (Claverie-Martin *et al.*, 1991). Furthermore, in the C-terminal half, which appears to constitute a separate domain, a shorter region of 86 amino acid residues is 29% homologous with the highly conserved 70 kDa protein of the human U1 small nuclear ribonucleoprotein particle (snRNP) (Claverie-Martin *et al.*, 1991, Casaregola *et al.*, 1992). A comparison of the reports on the $ams^{ts}$ and $rne^{ts}$ phenotypes shows that the *ams/rne* gene product is involved in controlling the processing and degradation of a large number of RNAs (Table 2). As the *ams/me* mutations seem to inhibit the activity of RNase E both *in vivo* and *in vitro*, RNase E would appear to be the key enzyme controlling mRNA decay in *E. coli*. However, it has not yet been possible to obtain a homogeneously pure preparation of RNase E with which to verify that this enzyme in fact cuts at all the *ams/me*-sensitive sites identified *in vivo*. This caveat is important in light of the ability of *ams/me* mutations to inhibit RNA cleavage *in vivo* at some sites thought to be targets of other nucleases, such as RNase K, RNase F, RNase P, and RNase III (Pragai and Apirion, 1982; Gurewitz *et al.*, 1983; Srivastava *et al.*, 1990; Lundberg *et al.*, 1990).

**4. RNase E:** RNase E was originally defined as a ribonuclease that processes the 9S ribosomal RNA precursor to yield 5S ribosomal RNA (Misra and Apirion, 1980). Consequently 9S RNA processing has been mostly employed as assay in the attempts to isolate RNase E (Misra and Apirion, 1979; Roy and Apirion, 1983; Chauhan *et al.*, 1991). Previous reports of the molecular weight and subunit composition of RNase E vary; in one of the reports a size of 66 kDa was suggested (Misra and Apirion, 1979; Roy and Apirion, 1983; Chauhan *et al.*, 1991). The activity of RNase E seems to depend on $Mg^{2+}$ and on $K^+$ (Misra and Apirion, 1979). Since the activity of RNase E is thermosensitive in extracts prepared from the *mets* mutants (Misra and Apirion, 1980) and since RNase E activity is increased in extracts from cells with an amplified *ams/me* gene (Babitzke and Kushner, 1991), it is suggested that RNase E is identical with the *ams/me* gene product. Recent purification strategies are therefore designed to purify RNase E activity alongside with the intact *ams/me* gene product (Carpousis, 1993 and Carpousis, Personal communication). The usage of protease inhibitors made it indeed possible to obtain preparations with the intact *ams/me* as one major protein component where RNase E activity was dramatically increased (Carpousis, Personal communication, Carpousis, 1993). However, because of the other components that were found in this preparations, it cannot yet be concluded that the *ams/me* gene product is necessary and sufficient for RNase E activity. In this connection it is worth mentioning that an RNase E-like activity was also co-purified with the heat shock chaperonin GroEL and that this preparation contained two major protein components with sizes of 17 kDa and 66 kDa (Sohlberg et al., 1993).

**5. RNase K and the *ams/me* gene:** The phenotypic *in vivo* changes observed in the *ams/me* mutant cells, after shift to non-permissive temperature include a retardation of bulk mRNA decay, a decreased activity of RNase E, but also of RNase K, RNase F and RNase III (Pragai and Apirion 1982; Gurewitz et al., 1983; Srivastava et al., 1990, Lundberg et al., 1990). RNase K has been attributed to the cleavages that seem to regulate the stability of the *ompA* transcript in respect with growth rate changes (Melefors and von Gabain, 1988, Nilsson et al., 1998, Lundberg et al., 1990). The relationship of RNase E and RNase K is not yet clear: *In vitro* studies with extracts from *ams^ts* and *me^ts* mutant cells, show that RNase K (*ompA* mRNA cleavage) unlike RNase E (9S RNA cleavage) is not temperature-sensitive in the mutants' background (Nilsson et al., 1988, Harde et al.; submitted). This result was actually the driving force to attempt purifying an RNase K activity which is distinct from RNase E (Lundberg et al., 1990). While it was not possible to separate RNase E activity (9S cleavage) from RNase K activity (*ompA* cleavage), we isolated recently an RNase E-like activity that functionally interacted with the chaperonin GroEL and that had no detectable RNase K activity (Sohlberg et al., 1993). This result is remarkable in the light of the previous finding that the *ams^ts* mutation could be complemented by a truncated *groEL* gene (Chanda *et al.*, 1985). *In vivo* RNase E and RNase K activity are regulated in opposite manners in respect with growth rate changes (Georgellis et al., 1992). The differential behavior of the two nuclease activities in respect with growth rate and the apparent discrepancies of RNase K activity (if the *in vivo* and *in vitro* situations are compared in the *me^ts* mutant), motivated us to analyze the two nuclease activities in the two *ams/me* mutant strains at different growth phases. At all growth phases, RNase K activity seems to be only affected *in vivo* in mutant cells at non-permissive temperature, while in *in vitro* extracts from mutant cells, RNase K activity (*ompA* mRNA cleavage) is not thermo-sensitive (Harde et al., submitted). In contrast, RNase E activity seems to depend - *in vivo* and *in vitro* - on an intact *ams/me* gene product **only** during exponential growth. However, during early stationary phase RNase E activity (9S cleavage) is not at all affected by the muta-tions in the *ams/me* gene (Harde et al., submitted). The purest preparation of RNase K consisted of five major polypeptides of sizes ranging from 21 to 62 kDa, when analyzed on a silver-stained denaturing gel (Lundberg et al., 1990). It is not yet clear which of these polypeptides are needed for RNase K activity, which elutes as a 55-60 kDa enzyme upon gel filtration under nondenaturing conditions (Lundberg *et al.*, 1990). As a possible rationalization to reconcile the common and distinct features of the two endoribonucleases, it may be suggested that RNase K is a proteolytic fragment of RNase E. Another possibility is that the two enzymes may share a common core but differ with respect to accessory factors. Only the purification of RNase E and RNase K to homogeneity will clarify their relationship.

**6. *ams/me*-dependent cleavage sites:** To date, all RNA sites classified as targets for cleavage by RNase E or other *ams/me* controlled nucleases, such as RNase K, have been identified on the basis of *in vivo* analysis of *ams/me* mutants or *in vitro* assays with partially purified enzymes. This is important to bear in mind that because cleavage site has yet been defined using homogeneously pure enzyme, definitive assignment of these cleavages to a particular enzyme may be difficult (Table 2). Indeed, some of these sites are believed actually to be targets not of RNase E or RNase K, but of RNase III, RNase P, or RNase F (Pragai and Apirion, 1982; Gurewitz *et al.*, 1983; Srivastava *et al.*, 1990). Such ambiguity might result in cases where prior cleavage by RNase E or RNase K elsewhere in the RNA molecule must occur before another ribonuclease can recognize and cut the RNA at the site in question.

A comparison of *ams/rne*-dependent cleavage sites led Ehretsmann *et al.* (1992) to suggest RAUUW (R = G or A; W = U or A) as a consensus RNase E cleavage site. *In vivo* and *in vitro* studies of the effect of point mutations on RNase E cleavage of phage T4 gene 32 mRNA are consistent with this consensus sequence and also suggest that cleavage by RNase E may be enhanced by a downstream stem-loop flanking the cleavage site (Ehretsmann *et al.*, 1992). However, an investigation of RNase E cleavage of RNA I variants *in vivo* does not support the concept of a canonical RNase E cleavage sequence (Lin-Chao and Cohen, 1991b).

Recently, Bouvet and Belasco have obtained *in vivo* evidence that efficient endonucleolytic cleavage of RNA I by RNase E is facilitated by the presence of several unpaired nucleotides at the RNA 5′ end (Bouvet and Belasco, 1993). This apparent 5′-end dependence of RNase E could explain how 5′-terminal secondary structure can impede degradation of mRNA in *E. coli* (Emory *et al.*, 1992). *In vitro* analysis of the sites of cleavage by partially purified RNase K in the 5′ untranslated region of *ompA* mRNA has identified GYXUUU (Y = G, A, or U; X = A or C) as a possible RNase K consensus sequence (Lundberg, 1991). Secondary structure analysis of the *ompA* 5′ UTR suggests that these RNase K cleavage sites reside in regions that are not base-paired (Chen *et al.*, 1991; Rosenbaum *et al.* 1993).

Like RNase III, RNase E is both an anabolic and catabolic ribonuclease, and cleavage by this enzyme can lead to relatively stable upstream and downstream processing products or to rapid degradation of both cleavage products. There are even examples of *ams/rne*-dependent cleavages that initiate decay of the upstream but not the downstream cleavage product, and *vice versa* (Table 2). The structural basis for selective degradation of upstream cleavage products seems straightforward. These presumably are readily digested by 3′-to-5′ exonucleases (RNase II and polynucleotide phosphorylase) unless protected by a hairpin structure upstream of the cleavage site (Mott *et al.*, 1985; Donovan and Kushner, 1986; Plamann and Stauffer, 1990; McLaren *et al.*, 1991). As pointed out in the initial section of this article, it is more difficult to explain why downstream cleavage prod-

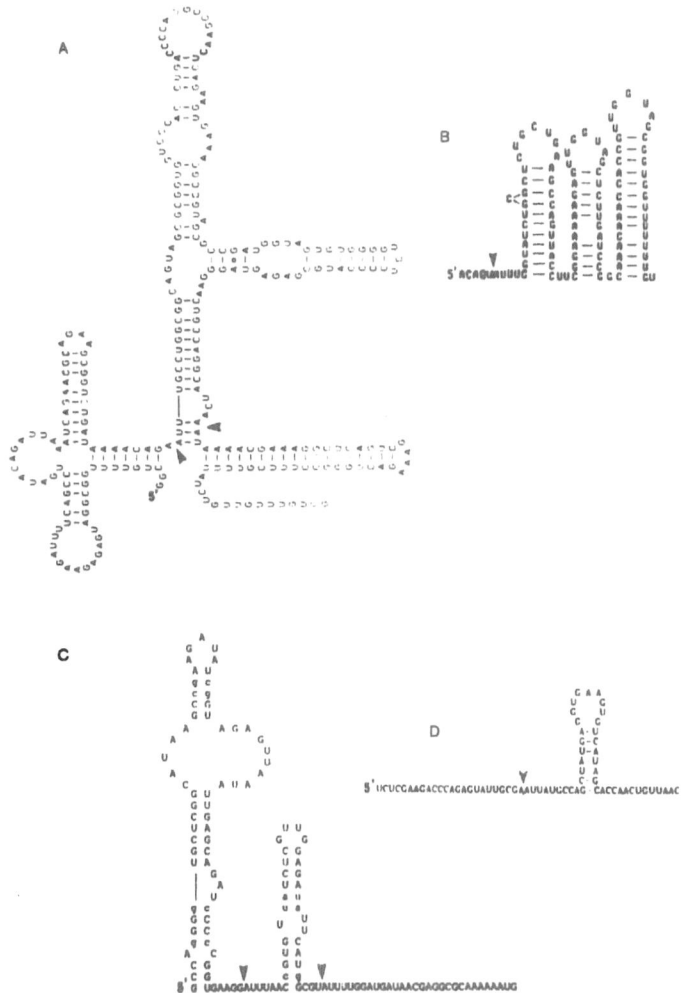

Fig. 10

RNase E and RNase K cleavage sites and RNA secondary structures. Figures 10A, 10B and 10D indicate the RNase E cleavage sites (arrows) relative to the published secondary structures of T4 gene32 RNA (Ehretsman *et al.*, 1992), RNA I (Lin-Chao and Cohen, 1991) and 9S RNA (Roy and Apirion, 1983, Christiansen, 1988), respectively. Figure 10C indicates the RNase K cleavage sites relative to the published secondary structure of the 5' UTR of the *ompA* mRNA (Chen *et al.*, 1991). The shown 5' UTR of the ompA mRNA is identical with one of the three co-existing structures that has been found in the study by Rosenbaum *et al.* (1993).

ducts are often susceptible to further rapid degradation, as no 5'-to-3' exoribonucleases have been found so far in bacteria (Deutscher and Zhang, 1990). For example, it has been shown that a downstream cleavage product that differs from a
primary transcript by only eight 5'-terminal nucleotides (i.e., the downstream
product of A-site cleavage of the *tacompA* transcript versus wild-type *ompA* mRNA
(Lundberg *et al.*, 1990)) or by as little as a 5' triphosphate versus a 5' monophosphate (i.e., the pRNA I-5 cleavage product of RNA I versus its triphosphate homolog
pppRNA I-5 (Lin-Chao and Cohen, 1991) can be much less stable despite the
seemingly minor structural difference. The mere absence of a 5'-terminal triphosphate is not adequate to explain the relatively rapid degradation of these downstream cleavage products, as other cleavage events result in 5'-monophosphory-
lated downstream products that are relatively stable (Table 2). Instead, cleavages
that initiate decay may be coupled mechanistically to the degradation of the 3'
cleavage product. For example, certain structural features downstream of a
cleavage site might facilitate or impede subsequent digestion by an as yet undis-
covered 5'-to-3' exonuclease or by an endonuclease that "scans" RNA in a 5' to 3'
direction.

## F. OTHER ENDONUCLEASES

**1. RNase I (Table 1):** RNase I is a broad specificity endoribonuclease that can cleave
rRNA between each residue to give mononucleotides. Under normal growth
conditions RNase I does not seem to participate in messenger turnover since global
mRNA degradation is not affected in RNase I deficient strains. It may be important
under some stress conditions (Spahr and Hollingworth, 1968; Neu and Heppel, 1964).
A cytoplasmic form of RNase I* has been purified (Cannistraro and Kennel, 1991).

**2. RNase M: (Table 1):** This enzyme has been purified *in vitro*. It has low specificity
and cleaves RNA between pyrimidine and adenosine residues. The similarity in
protein size, 24 KDa, and tryptic protein digestion pattern between RNase I and M
indicate that these two enzymes are closely related (Meador et al., 1990). Both
RNases are coded by the same gene, *rna*. They may differ in their post-translational
modification. Two other activities RNase IV and F also have some characteristics
rather similar to those of RNases M and I suggesting that all four of these activities are
related but their role in mRNA decay is undetermined. RNase N is a partially purified
activity that cleaves various double and single stranded RNA substrates non
specifically to give mononucleotides (Misra and Apirion, 1978).

**3. Reg B RNase:** This endoribonuclease is encoded by the T4 genome; it cleaves early mRNA of phage T4 and probably host mRNAs at the beginning of phage infection but it does not cleave middle and late mRNAs of phage T4 (Uzan et al., 1988). The enzyme is apparently only active at the beginning of T4 infection (Sanson and Uzan, 1993). The enzyme is an 18 KDa polypeptide. It cleaves specifically on the sequence 5'GGAG3' between the G and A to give GpGp and ApG with a free 5'OH group. The cleavage site frequently occurs within the Shine and Dalgarno sequence. In this case there is a strong inhibition of protein synthesis. When the site is not within a Shine and Dalgarno sequence there is a large decrease in the chemical half life of the mRNA and a small decrease in the functional mRNA half life. In conclusion the RegB RNase has two roles : 1) inhibition of a class of early genes which have a GGAG site in the Shine and Dalgarno sequence. 2) general decrease of mRNA half-life at the beginning of infection (early phage mRNAs and host mRNAs).

ACKNOWLEDGEMENT

We are grateful to Anette Hedberg and Petra Rosén for helping to prepare the manuscript. This article was supported by grants to A von Gabain from the Swedish Cancer Society (RmC) and from the Federal Repulic of Austria.

**REFERENCES**

Apirion, D. and Watson M. (1975) J Bact 124: 317-324

Apirion, D. (1978). Genetics 90: 659-671

Babitzke, P. and Kushner, S.R. (1991) Proc Natl Acad Sci USA 88: 1-5

Bardwell, J.C.A., Regnier P., Chen S.M., Nakamura, Y., Grunberg-Manago, M. and Court D.L. (1989) EMBO J. 8: 3401-3407

Baumeister, R., Flache, P., Melefors, Ö., von Gabain, A. and Hillen, W. (1991) Nucleic Acids Res 19: 4595-4600

Beltchev, B., Thang, M.N. and Portier, C. (1971) Eur J Biochem 19: 194-199

Bouvet, P. and Belasco, J.G (1992) Nature 360: 488-491

Brun, Y.V., Sanfacon, H., Breton, R. and Lapointe, J. (1990) J Mol Biol 214: 845-864

Cannistraro V.J. and Kennell D. (1989) Eur J Biochem 181: 363-370

Cannistraro V.J. and Kennell D. (1991) J Bacteriol 173: 4653-4659

Carpousis, A.J., Mudd, E.A., and Krisch, H.M. (1989) Mol Gen Genet 219: 39-48

Carpousis, A.J (1993) Presented at the Keystone meeting on "Nucleases"

Casaregola, S., Norris, V., Goldberg, M. and Holland, I. B. (1990) Mol Microbiol 4: 505-511

Casaregola S., Jacq, A. Laoudj, D. McGurk G., Margason G., Tempete M., Norris V. and Holland I.B. (1992) J Mol Biol 228: 30-40

Chelladurai, B.S., Li, H. and Nicholson, A.M. (1991) Nucleic Acids Research 19: 1759-1766

Chanda, P. K., Ono, M., Kuwano, M., and Kung, H.-F. (1985) J Bacteriol 161: 446-449

Chauhan, A.K and Apirion, D. (1991) Mol Gen Genet 228: 49-54

Chauhan, A.K., Miczak, A., Taraseviciene, L. and Apirion, D. (1991) Nucleic Acids Res 19: 125-129

Chen, L.-H., Emory., S.A., Bricker, A. L., Bouvet, P. and Belasco, J. (1991) J Bacteriol 173: 4578-4586

Claverie-Martin, F., Diaz-Torres, M.R., Yancey, S.D. and Kushner, S.R. (1989) J Bacteriol 171: 5479-5486

Claverie-Martin, F., Diaz-Torres, M.R., Yancey, S.D. and Kushner, S.R. (1991) J Biol Chem 266: 2843-2851

Christiansen, J. (1988) Nucleic Acid Res 16: 7457-7476

Court, D. (1993) in "Control of mRNA stability" (eds. Brawerman, G and Belasco, J) Academic Press

Dallmann, G., Dallmann, K., Sonin, A., Miczak, A. and Apirion, D. (1987) Mol Gen (Life Sci Adv) 6: 99-107

Deutscher, M.P. (1985) Cell 40: 731-732

Deutscher, M.P. and Zhang, J. (1990) (J.E.G. McCarthy and M.F. Tuite, eds) NATO ASI Series Vol. 49: pp 1-12. Springer-Verlag Berlin Heidelberg

Deutscher, M.P. and Reuven, B (1991). Proc Natl Acad Sci USA 88: 3277-3280

Donovan, W.P., and Kushner, S.R. (1986) Proc Natl Acad Sci USA 83: 120-124

Dunn, J.J. and Studier, F.W. (1973) Proc Natl Acad Sci USA, 70: 3296-3300

Dunn, J.J. and Studier, F.W. (1975) J Mol Biol 99: 487-499

Dunn, J.J. (1976) J Mol Biol 99: 487-499

Ehretsmann, C.P., Carpousis, A. J. and Krisch, H. M. (1992) Genes Dev 6: 149-159

Emory et al., (1992) Genes Dev 6: 135-148

Faubladier, M., Cam, K. and Bouche, J.P. (1990) J Mol Biol 212: 461-471

Georgellis, D., Arvidsson, S. and von Gabain, A. (1992) J Bact 174: 5382-5390.

Ghora, B.K., and Apirion, D. (1978) Cell 15: 1055-1066

Gitelman, D.R. and Apirion D. (1980) Bioch Biophys Res Commun 96: 1063-1070

Godefroy, T. (1970) Eur J Biochm 14: 222-231

Godefroy-Colburn T. and Grunberg-Manago M. (1972) "The enzymes" v. 7 (3ème ed.). Acad. Press N.Y., pp.533-583

Goldblum, K and Apirion, D. (1981) J Bact 146: 128-132

Gross, G. (1991) J Biol Chem (1991) 10: 17885-17889

Guarneros, G. (1988) "Current topics in Microbiology and Immunology" Hobom G. Ron R. eds, Springer Verlag, Berlin pp. 1-19

Guarneros, G. and Portier, C. (1990) Biochimie 72: 771-777

Gupta, R.S., Kasai, T. and Schlessingen, D. (1977) J Biol Chem 252: 8945-8948

Gurewitz, M., Watson, M. and Apirion, D. (1982) Eur J Biochem 124: 553-559

Gurewitz, M., Jain, S.K. and Apirion, D. (1983) Proc Natl Acad Sci USA 80: 4450-4454

Harde, T., Sohlberg, B., Georgellis, D., Melefors, Ö., Lundberg, U. and von Gabain, A. Submitted for publication

Higgins, C.F., Mc Laren, RS and Newbury, S.F. (1988) Gene 72: 3-14

Kokoska, R.J., Blumer, K.J. and Steege, D.A. (1990) Biochimie 72: 803-811

Kameyama, L., Fernandez, L., Court, D. and Guarneros, G. (1991) Mol Microbiol 5: 2953-2963

Krinke, L. and Wulff, D.L. (1987) Genes Dev 1: 1005-1013

Krinke, L. and Wulff, D.L. (1990) Nucleic Acid Res 18: 4809-4815

Kuwano, M., Ono, M., Endo, H., Hori, K., Nakamura, K., Hirota, Y. and Ohnishi, Y. (1977) Mol gen Genet 154: 279-285

Lin-Chao, S. and Cohen, S.N. (1991a) Cell 65: 1233-1242

Lin-Chao, S. and Cohen, S.N. (1991b) presented at the ASM meeting on mRNA decay and RNA processing

Lundberg, U., von Gabain, A. and Melefors, Ö. (1990) EMBO J 9: 2731-2741

Lundberg, U. (1991) Thesis, Karolinska Institute, Stockholm

Mackie, G.A. (1991) J Bact 173: 2488-2497

McDowall, K.J., Hernandez, R.G., Lin-Chao, S. and Cohen, S.N. (1993). J Bact In press

McLarens, R. F., Newbury, S. F., Dance, G. S. C., Gauston, H.C. and Higgins, C. (1991) J Mol Biol 221: 81-95

Meader, J., Cannon, B., Cannistraro, V.J. and Kennel, D. (1990) Eur J Biochem 187: 549-553

Melefors, Ö. (1991) Thesis, Karolinska Institute, Stockholm

Melefors, Ö. and von Gabain, A. (1988) Cell 52: 893-901

Melefors, Ö. and von Gabain, A. (1991) Mol Microbiol 5: 857-864

Misra, T.K. and Apirion D. (1979) J Biol Chem 254: 11154-11159

Misra, T.K. and Apirion D. (1980) J Bact 142: 359-361

Mott, J. E., Galloway, J. L. and Platt, T. (1985) EMBO J 4: 1887-1891

Mudd, E.A., Prentki, P., Belin, D. and Krisch, H.M. (1988) EMBO J 7: 3601-3607

Mudd, E.A., Carpousis, A.J. and Krisch, H.M. (1990a) Genes and Dev 4: 873-881

Mudd, E.A., Krisch, H.M and Higgins, C.F. (1990b) Mol Micriobiol 4: 2127-2135

Neu, H.C. and Heppel, L.A. (1964) Res Commun 14: 109-112

Nilsson, G., Belasco, J., Cohen, S. N. and von Gabain, A. (1984) Nature 312: 75-77

Nilsson, G., Lundberg, U., and von Gabain, A. (1988) EMBO J 7: 2269-2275

Nilsson, P. and Uhlin, B.E. (1991) Molec Micriobiol 5: 1791-1799

Nossal, N.G. and Singer, M.F. (1968) J Biol Chem 243: 913-922

Ono, M and Kuwano, M. (1979) J Mol Biol 129: 343-357

Plamann, M.D. and Stauffer, G.V. (1990) Mol Gen Genet 220: 301-306

Portier, C. (1975) Eur. J. Biochem 55: 573-583

Portier, C., Dondon L., Grunberg-Manago M. and Regnier Ph. (1987) EMBO J 6: 2165-2170

Pragai, B. and Apirion, D. (1982) J Mol Biol 154: 465-484

Regnier, P., Grunberg-Manago, M. and Portier C. (1987) J Biol Chem 262: 63-70

Regnier, P., and Grunberg-Manago M. (1990) Biochimie 72: 825-834

Regnier, P., and Hajnsdorf, E. (1991). J Mol Biol 217: 283-292

Robert-Le Meur, M. and Portier, C. (1992) EMBO J 11: 2633-2641

Robertson, H.D., Webster, R.E. and Zinder, N.D. (1968) J Biol Chem 243: 82-91

Rosenbaum, V., Klahn, T., Lundberg, Holmgren, E., von Gabain, A. and Riesner, D. (1993) J Mol Biol 229: 656-670

Roy, M.K. and Apirion D. (1983) Biochimica et Biophysica Acta 747: 200-208

Sanson, B. and Uzan, M. (1993) J Mol Biol submitted

Sohlberg, B., Lundberg, U., Hartl, U. and von Gabain, A. Proc Nat Acad Sci 90: 277-281

Spahr, P.F. and Gesteland, R.F. (1968) Proc Natl Acad Sci USA 59: 876-883

Spahr, P.F. and Hollingworth, B.R. (1961) J Biol Chem 236: 823-831

Srivastava, R.K., Miczak, A., and Apirion, D. (1990) Biochimie 72: 791-802

Takiff, H.E., Chen, S.M. and Court, D.L. (1989) J Bact 171: 2581-2590

Tarasiviciene, L., Miczak, A. and Apirion, D (1991) Mol Microbiol 5: 851-855

Thang, M.N., Guschbauer, W., Zachan, H.G. and Grunberg-Manago, M. (1967) J Mol
     Biol 26: 403-421

Tomcsanyi, T. and Apirion, D. (1985) J Mol Biol 185: 713-720

Uzan, M. Favre, R. and Brody, E. (1988) Proc Natl Acad Sci USA 85: 8895-8899

Zilhao, R., Camelo, L. and Arraiano, C.M. (1993) Mol Microbiol. In press

# Mechanisms and Structural Determinants of mRNA Decay in Bacteria

Joel G. Belasco
Department of Microbiology and Molecular Genetics
Harvard Medical School
200 Longwood Avenue
Boston, Massachusetts 02115, USA

Messenger RNA degradation is an important mechanism for controlling gene expression in all organisms. Recent investigations have shed light on the mechanisms of mRNA decay in bacteria and the elements in prokaryotic mRNAs that are responsible for their marked differences in longevity. These studies have identified several ribonucleases that participate in mRNA degradation and have shown that specific structural features near the 5' and 3' ends of mRNA can play a key role in determining mRNA longevity in vivo.

## I. Structural barriers to 3'-exonucleolytic attack

Despite their importance for the complete degradation of mRNA fragments to mononucleotides, it appears that 3' exoribonucleases (RNase II, polynucleotide phosphorylase) do not control the rate-determining step in the decay of most intact mRNAs in *Escherichia coli* and other bacteria (Donovan and Kushner, 1983; Donovan and Kushner, 1986; Chen et al., 1988). This apparently is due to the fact that almost all detectable prokaryotic mRNAs end with a 3' stem-loop structure, which can severely impede the initiation or propagation of 3'-exonuclease digestion and thereby force degradation to begin with endonucleolytic cleavage at an upstream site (Mott et al., 1985; Newbury et al., 1987; Chen et al., 1988; Plamann and Stauffer, 1990; Chen et al., 1991; McLaren et al., 1991). This important but limited role of stem-loop structures in protecting mRNA from degradation is nicely illustrated by the decay of *puf* mRNA in the photosynthetic bacterium *Rhodobacter capsulatus*.

The *puf* operon (*pufQBALMX*) of *R. capsulatus* encodes several proteins, four of which (the *pufB, pufA, pufL*, and *pufM* gene products) participate in the initial events of photosynthesis (Youvan et al., 1984; Belasco et al., 1985).

NATO ASI Series, Vol. H 97
Post-transcriptional Control of Gene Expression
Edited by Orna Resnekov and Alexander von Gabain
© Springer-Verlag Berlin Heidelberg 1996

Interestingly, despite being cotranscribed as part of a single operon, the *pufB* and *pufA* genes are expressed at a ~12-fold higher level than the *pufL* and *pufM* genes (Drews, 1985). The basis for this differential gene expression was clarified when the two principal *puf* mRNAs in *R. capsulatus* were identified and quantified. One is a 2.7 kb *pufBALMX* mRNA; the other, which is 9-fold more abundant, is a 0.5 kb *pufBA* mRNA (Belasco et al., 1985). The greater abundance of *pufBA* mRNA, which can account for the higher level of expression of the *pufB* and *pufA* genes, is due principally to the longevity of this 0.5 kb mRNA in vivo. The half-life of *pufBA* mRNA is 20 min, versus 3 min for *pufBALMX* mRNA (Belasco et al., 1985). Pulse-chase labeling experiments have shown that *pufBA* mRNA arises primarily as a long-lived intermediate in the decay of *pufBALMX* mRNA (Belasco et al., 1985).

The greater stability of the 5' *pufBA* segment of *pufBALMX* mRNA is due in part to the presence of a large stem-loop structure at the boundary between the *pufBA* and *pufLMX* segments. This stem-loop apparently blocks 3'-to-5' exonucleolytic propagation of decay from the labile *pufLMX* segment into the comparatively stable *pufBA* segment (Belasco et al., 1985; Chen et al., 1988). Consequently, deletion of this intercistronic hairpin structure reduces the half-life of the *pufBA* segment to value similar to that of the *pufLMX* segment (Chen et al., 1988; Belasco and Chen, 1988). Despite the importance of this stem-loop, the difference in stability of the 5' and 3' segments of *pufBALMX* mRNA is not due solely to the ability of this RNA hairpin to resist 3' exonuclease attack. The *pufLMX* segment itself ends with a pair of 3' stem-loops, and a variety of experiments indicate that degradation of this short-lived segment begins with rate-determining endonucleolytic cleavage at an internal site rather than with 3'-exonucleolytic penetration of these terminal stem-loops (Chen et al., 1988; Belasco and Chen, 1988; Chen and Belasco, 1990). Therefore, the greater longevity of the 5' *pufBA* segment is due also to its greater resistance to endonucleolytic cleavage.

## II. Endonucleolytic initiation of mRNA decay

To date, only three *E. coli* endonucleases have been shown to initiate mRNA decay in vivo: RNase III, RNase E and RNase K. Cleavage by RNase III is the first step in the degradation of some bacterial and phage mRNAs, including the *E. coli* transcripts that encode RNase III and polynucleotide phosphorylase and the antiterminated lambda $P_L$ transcript (Schmeissner et al., 1984; Portier et al., 1987; Bardwell et al., 1989). However, this endonuclease does not participate in the degradation of most *E. coli* mRNAs (Apirion and Watson, 1975). RNase E cleavage appears to trigger the decay of a much larger number of prokaryotic RNAs, including RNA I of plasmid pBR322 (a labile untranslated RNA that controls plasmid copy number) and many mRNAs (Lin-Chao and Cohen, 1991). Thermal inactivation of a gene product necessary for RNase E activity also increases the average half-life of bulk mRNA in *E. coli*, suggesting that cleavage by RNase E may be the initial step that controls the decay rate of most *E. coli* mRNAs (Apirion, 1978; Ono and Kuwano, 1979; Mudd et al., 1990; Babitzke and Kushner, 1991; Melefors and von Gabain, 1991; Taraseviciene et al., 1991). RNase K has been implicated in the accelerated decay of *E. coli ompA* mRNA under conditions of slow bacterial growth (Melefors and von Gabain, 1988; Lundberg et al., 1990).

### III. 5' mRNA stabilizers

The 5' leader segments of certain long-lived prokaryotic mRNAs have been shown to function as mRNA stabilizers that can prolong the lifetime of heterologous mRNAs to which they are fused (Yamamoto and Imamoto, 1975; Gorski et al., 1985; Belasco et al., 1986; Bechhofer and Dubnau, 1987; Sandler and Weisblum, 1988). Among these, the best studied are the 5' untranslated region (UTR) of the *E. coli ompA* transcript and the 5' leader of the *Staphylococcus aureus ermC* and *ermA* mRNAs.

The *ompA* transcript, which encodes an abundant outer membrane protein important for bacterial conjugation, is one of the most stable mRNAs in *E. coli*. Its unusual longevity in rapidly growing cells (half-life = 17 min) is due in large part to its 5' UTR, which can stabilize a variety of mRNAs to which it is fused (Belasco et al., 1986; Emory and Belasco, 1990; Hansen et al., 1993). Secondary-structure analysis has revealed that this 5' UTR comprises four domains: a 5'-terminal stem-loop (hp1), an internal stem-loop (hp2), a single-stranded segment between the two stem-loops (ss1), and a second single-stranded segment (ss2) that includes the ribosome binding site and flanking sequences and that is located downstream of hp2 (Chen et al., 1991). Of these four domains, only hp1 and ss2 are important for the long lifetime of the *ompA* transcript (Emory et al., 1992). Interestingly, it is the 5'-terminal location of hp1, rather than its sequence or shape, that is most critical for its function. In fact, the addition of several unpaired nucleotides upstream of hp1 is as destabilizing as the deletion of this hairpin (Emory et al., 1992).

Recent experiments with pBR322 RNA I suggest that the role of the 5'-terminal *ompA* stem-loop may be to impede downstream cleavage of mRNA by RNase E (Bouvet and Belasco, 1992). Degradation of RNA I is triggered by RNase E cleavage at a site five nucleotides from the 5' end (Tomcsanyi and Apirion, 1985; Lin-Chao and Cohen, 1991). Addition of a simple stem-loop to the 5' end of RNA I can slow cleavage at this site by as much as a factor of five (Bouvet and Belasco, 1992). Rapid RNase E cleavage is restored by adding five unpaired nucleotides upstream of the stem-loop (Bouvet and Belasco, 1992). Therefore, it appears that a 5'-terminal stem-loop can impair the ability of RNase E to gain access to an internal cleavage site. This finding suggests that the mechanism of action of RNase E may have an intrinsic 5'-to-3' directionality.

A similar directionality of mRNA decay is evident from studies of mRNA stabilization in *Bacillus subtilis* by the leader segment of *ermC* and *ermA* mRNA. These transcripts, which encode resistance to erythromycin, are inducibly stabilized by low doses of this antibiotic (Bechhofer and Dubnau, 1987; Sandler and Weisblum, 1988). In each case, the mechanism of stabilization involves erythromycin-induced ribosome stalling at a sensitive site within a short translational open reading frame in the 5' leader segment. Experiments with hybrid mRNAs indicate that the stalled ribosome protects downstream, but not upstream, sequences from degradation (Bechhofer and Zen, 1989; Sandler and Weisblum, 1989). This finding suggests that an important pathway for mRNA decay in *B. subtilis* involves a ribonuclease (RNase E? a 5' exonuclease?) that degrades mRNA from 5' to 3'.

## IV. Propagation of decay

In most cases, endonucleolytic cleavage of prokaryotic mRNA triggers very rapid digestion of the resulting 5' and 3' fragments. The lability of the 5' fragment can be explained by the action of bacterial 3' exonucleases, which will readily degrade virtually any RNA that lacks an untranslated 3'-terminal stem-loop (Mott et al., 1985; Newbury et al., 1987; Chen et al., 1988; Plamann and Stauffer, 1990; Chen and Belasco, 1990; Chen et al., 1991; McLaren et al., 1991). The reason for the frequent lability of the 3' cleavage product is less apparent, but in principal could be due to the action of a hypothetical 5' exonuclease that preferentially degrades the 3' products of endonucleolytic cleavage or an endonuclease that acts processively to degrade RNA from 5' to 3'. Alternatively, swift degradation of the 3' fragment could be facilitated by a loss of associated ribosomes if the initial cleavage event severs an upstream signal for translation initiation.

## References

Apirion, D. (1978). Isolation, genetic mapping, and some characterization of a mutation in *Escherichia coli* that affects the processing of ribonucleic acids. Genetics 90, 659-671.

Apirion, D., & Watson, N. (1975). Unaltered stability of newly synthesized RNA in strains of *Escherichia coli* missing a ribonuclease specific for double-stranded RNA. Mol. Gen. Genet. 136, 317-326.

Babitzke, P., & Kushner, S. R. (1991). The Ams (altered mRNA stability) protein and ribonuclease E are encoded by the same structural gene of *Escherichia coli*. Proc. Natl. Acad. Sci. USA 88, 1-5.

Bardwell, J. C. A., Régnier, P., Chen, S.-M., Nakamura, Y., Grunberg-Manago, M., & Court, D. L. (1989). Autoregulation of RNase III operon by mRNA processing. EMBO J. 8, 3401-3407.

Bechhofer, D. H., & Dubnau, D. (1987). Induced mRNA stability in *Bacillus subtilis*. Proc. Natl. Acad. Sci. USA 84, 498-502.

Bechhofer, D. H., & Zen, K. H. (1989). Mechanism of erythromycin-induced *ermC* mRNA stability in *Bacillus subtilis*. J. Bacteriol. 171, 5803-5811.

Belasco, J. G., Beatty, J. T., Adams, C. W., von Gabain, A., & Cohen, S. N. (1985). Differential expression of photosynthesis genes in *R. capsulata* results from segmental differences in stability within the polycistronic *rxc*A transcript. Cell 40, 171-181.

Belasco, J. G., & Chen, C.-Y. A. (1988). Mechanism of *puf* mRNA degradation: the role of an intercistronic stem-loop structure. Gene 72, 109-117.

Belasco, J. G., Nilsson, G., von Gabain, A., & Cohen, S. N. (1986). The stability of *E. coli* gene transcripts is dependent on determinants localized to specific mRNA segments. Cell 46, 245-251.

Bouvet, P., & Belasco, J. G. (1992). Control of RNase E-mediated RNA degradation by 5'-terminal base pairing in *E. coli*. Nature 360, 488-491.

Chen, C.-Y. A., Beatty, J. T., Cohen, S. N., & Belasco, J. G. (1988). An intercistronic stem-loop structure functions as an mRNA decay terminator necessary but insufficient for *puf* mRNA stability. Cell 52, 609-619.

Chen, C.-Y. A., & Belasco, J. G. (1990). Degradation of *pufLMX* mRNA in *Rhodobacter capsulatus* is initiated by non-random endonucleotyic cleavage. J. Bacteriol. 172, 4578-4586.

Chen, L.-H., Emory, S. A., Bricker, A. L., Bouvet, P., & Belasco, J. G. (1991). Structure and function of a bacterial mRNA stabilizer: Analysis of the 5' untranslated region of *ompA* mRNA. J. Bacteriol. 173, 4578-4586.

Donovan, W. P., & Kushner, S. R. (1983). Amplification of ribonuclease II (*rnb*) activity in *Escherichia coli* K-12. Nucleic Acids Res. 11, 265-275.

Donovan, W. P., & Kushner, S. R. (1986). Polynucleotide phosphorylase and ribonuclease II are required for cell viability and mRNA turnover in *Escherichia coli* K-12. Proc. Natl. Acad. Sci. USA 83, 120-124.

Drews, G. (1985). Structure and functional organization of light-harvesting complexes and photochemical reaction centers in membranes of phototrophic bacteria. Microbiol. Rev. 49, 59-70.

Emory, S. A., & Belasco, J. G. (1990). The *ompA* 5' untranslated RNA segment functions in *Escherichia coli* as a growth-rate-regulated mRNA stabilizer whose activity is unrelated to translational efficiency. J. Bacteriol. 172, 4472-4481.

Emory, S. A., Bouvet, P., & Belasco, J. G. (1992). A 5'-terminal stem-loop structure can stabilize mRNA in *Escherichia coli*. Genes Dev. 6, 135-148.

Gorski, K., Roch, J.-M., Prentki, P., & Krisch, H. M. (1985). The stability of bacteriophage T4 gene 32 mRNA: a 5' leader sequence that can stabilize mRNA transcripts. Cell 43, 461-469.

Hansen, M. J., Schoenberg, M. L., & Belasco, J. G. (1993). The *ompA* 5' untranslated region impedes a major pathway for mRNA degradation in *Escherichia coli*. (submitted).

Lin-Chao, S., & Cohen, S. N. (1991). The rate of processing and degradation of antisense RNA I regulates the replication of Col E1-type plasmids *in vivo*. Cell 65, 1233-1242.

Lundberg, U., von Gabain, A., & Melefors, O. (1990). Cleavages in the 5' region of the *ompA* and *bla* mRNA control stability: Studies with an *E. coli* mutant altering mRNA stability and a novel endoribonuclease. EMBO J. 9, 2731-2741.

McLaren, R. S., Newbury, S. F., Dance, G. S. C., Causton, H. C., & Higgins, C. F. (1991). mRNA degradation by processive 3'-5' exoribonucleases *in vitro* and the implications for prokaryotic mRNA decay *in vivo*. J. Mol. Biol. 221, 81-95.

Melefors, O., & von Gabain, A. (1988). Site-specific endonucleolytic cleavages and the regulation of stability of *E. coli ompA* mRNA. Cell 52, 893-901.

Melefors, O., & von Gabain, A. (1991). Genetic studies of cleavage-initiated mRNA decay and processing of ribosomal 9S RNA show that the *Escherichia coli ams* and *rne* loci are the same. Mol. Microbiol. 5, 857-864.

Mott, J. E., Galloway, J. L., & Platt, T. (1985). Maturation of *Escherichia coli* tryptophan operon mRNA: evidence for 3' exonucleolytic processing after rho-dependent termination. EMBO J. 4, 1887-1891.

Mudd, E. A., Krisch, H. M., & Higgins, C. F. (1990). RNase E, an endoribonuclease, has a general role in the chemical decay of *E. coli* mRNA: Evidence that *rne* and *ams* are the same genetic locus. Mol. Microbiol. 4, 2127-2135.

Newbury, S. F., Smith, N. H., Robinson, E. C., Hiles, I. D., & Higgins, C. F. (1987). Stabilization of translationally active mRNA by prokaryotic REP sequences. Cell 48, 297-310.

Ono, M., & Kuwano, M. (1979). A conditional lethal mutation in an *Escherichia coli* strain with a longer chemical lifetime of mRNA. J. Mol. Biol. 129, 343-357.

Plamann, M. D., & Stauffer, G. V. (1990). *Escherichia coli glyA* mRNA decay: the role of secondary structure and the effects of the *pnp* and *rnb* mutations. Mol. Gen. Genet. 220, 301-306.

Portier, C., Dondon, L., Grunberg-Manago, M., & Regnier, P. (1987). The first step in functional inactivation of the *Escherichia coli* polynucleotide phosphorylase messenger is a ribonuclease III processing at the 5' end. EMBO J. 6, 2165-2170.

Sandler, P., & Weisblum, B. (1988). Erythromycin-induced stabilization of *ermA* messenger RNA in *Staphylococcus aureus* and *Bacillus subtilis*. J. Mol. Biol. 203, 905-915.

Sandler, P., & Weisblum, B. (1989). Erythromycin-induced ribosome stall in the *ermA* leader: a barricade to 5'-to-3' nucleolytic cleavage of the *ermA* transcript. J. Bacteriol. 171, 6680-6688.

Schmeissner, U., McKenney, K., Rosenberg, M., & Court, D. (1984). Removal of a terminator structure by RNA processing regulates *int* gene expression. J. Mol. Biol. 176, 39-53.

Taraseviciene, L., Miczak, A., & Apirion, D. (1991). The gene specifying RNase E (*rne*) and a gene affecting mRNA stability (*ams*) are the same gene. Mol. Microbiol. 5, 851-855.

Tomcsanyi, T., & Apirion, D. (1985). Processing enzyme ribonuclease E specifically cleaves RNA I, an inhibitor of primer formation in plasmid DNA synthesis. J. Mol. Biol. 185, 713-720.

Yamamoto, T., & Imamoto, F. (1975). Differential stability of *trp* messenger RNA synthesized originating at the *trp* promoter and $p_L$ promoter of lambda *trp* phage. J. Mol. Biol. 92, 289-309.

Youvan, D. C., Bylina, E. J., Alberti, M., Begusch, H., & Hearst, J. E. (1984). Nucleotide and deduced polypeptide sequences of the photosynthetic rection-center, B870 antenna, and flanking polypeptides from *R. capsulata*. Cell 37, 949-957.

# CHANGES IN THE LENGTH OF POLY(A) TAILS AND THEIR EFFECTS ON mRNA TRANSLATION AND TURNOVER

Marvin Wickens[1], Joel G. Belasco[2], and Allan Jacobson[3]

[1]Department of Biochemistry
University of Wisconsin
420 Henry Mall
Madison, WI 53706
[2]Department of Microbiology and Molecular Genetics
Harvard Medical School
25 Shattuck Street
Boston, MA 02115
[3]Department of Molecular Genetics and Microbiology
University of Massachusetts Medical School
55 Lake Avenue North
Worcester, MA 01655

## Introduction

Most eukaryotic mRNAs have a sequence of polyadenylic acid [poly(A)] at their 3'-termini. These poly(A) tails are normally added post-transcriptionally in the nucleus with an initial length, in higher eukaryotes, of approximately 200-250 adenylate residues (Munroe and and Jacobson, 1990a; Wickens, 1990; Wahle and Keller, in press). Following transport of mRNA to the cytoplasm, poly(A) tracts are gradually shortened to lengths of approximately 30-60 adenylate residues and then, in many instances, are completely removed. Recent experiments, representing the convergence of work using very different organisms -- frog oocytes, yeast, cultured mammalian and plant cells -- indicate that these and other poly(A) addition and removal reactions may govern the translatability and/or turnover of specific mRNAs.

## Poly(A) addition and removal in oocytes and embryos

Although polyadenylation is normally a nuclear event, poly(A) can also be added (and removed) in the cytoplasm. Such adenylation and deadenylation has been observed in a variety of cells types, but most notably in oocytes and early embryos. Changes in poly(A) length during early development are often correlated with changes in translational activity, i.e., certain maternal mRNAs acquire poly(A) in the cytoplasm and concomitantly are translationally activated, while others lose it and

NATO ASI Series, Vol. H 97
Post-transcriptional Control of Gene Expression
Edited by Orna Resnekov and Alexander von Gabain
© Springer-Verlag Berlin Heidelberg 1996

are repressed (Munroe and Jacobson, 1990a; Wickens, 1992). Recent experiments have focused on the period of development termed oocyte maturation, in which oocytes advance from first to second meiosis. In both frog and mouse oocytes, this process can be induced in culture by progesterone, and is accompanied by the addition and removal of poly(A) from specific maternal mRNAs. In the generic experiments examining these reactions, synthetic maternal mRNAs, or portions thereof, are injected into the cytoplasm of the oocyte and the behavior of that RNA is then examined after inducing maturation. Through these conceptually simple experiments, several key features of the two reactions have already emerged.

Polyadenylation during oocyte maturation requires specific sequences in the mRNA 3'-UTR. One required element is the sequence AAUAAA, contained in all but certain histone mRNAs (reviewed in Wickens, 1992). A second required element appears, in some mRNAs, to be UUUUUAU or a close relative (Fox et al., 1989; McGrew et al., 1989). Removal of this sequence from a maternal mRNA can inactivate its polyadenylation, while insertion of this sequence into a previously inactive RNA can turn it on. Although UUUUUAU (combined with the AAUAAA sequence) is sufficient to signal poly (A) addition, it remains to be determined whether it is the best sequence, the shortest sequence, or the only sequence to do so. Indeed, preliminary results suggest that the elements required for polyadenylation during maturation may be quite plastic in sequence, flexible in position, and functionally redundant (Thompson and Wickens, unpublished exps.).

In contrast to poly(A) addition, poly(A) removal during maturation is a default reaction, i.e., it proceeds in the absence any specific sequences (Fox and Wickens, 1990; Varnum and Wormington, 1990). Poly(A) removal appears to be prevented only by poly(A) addition. One illustration of this point comes from the analysis of an RNA containing a poly(A) tail, but lacking the signals required for addition because of a point mutation in AAUAAA. During maturation, this RNA is not only inactive in polyadenylation, but it also loses its poly(A) tail precisely (Fox and Wickens, 1990).

While such experiments demonstrate that poly(A) removal does not require specific sequences, they do not address the question of whether specific sequences can affect the rate of the reaction, or determine its timing or extent. Indeed specific sequences in the 3'-UTRs of certain frog mRNAs (e.g., Eg1, Eg2 and Eg5) cause deadenylation to occur after fertilization (Bouvet et al., 1991). In addition, other mRNAs lose most of their tails in the oocyte before maturation, via a reaction that requires the same sequences in their 3'-UTRs that will be required later for them to receive poly(A) during maturation (Huarte et al., 1992). This seems to provide an elegant simplicity of design since the same sequence element first prevents translation of the mRNA by reducing its tail length, then later activates it by causing its tail to be extended. Whether the same factors are involved in recognizing the sequence in these two reactions is not yet known.

## The enzymes involved in poly (A) addition

Poly(A) addition in a crude egg extract requires the same sequences as it does *in vivo*, and by that criterion, accurately mimics the reaction in the cell (Paris et al., 1991; Fox et al., 1992). By fractionating the extract, two factors have been obtained that are essential. One contains a poly(A) polymerase that lacks intrinsic sequence specificity, while the other contains a sequence-specific RNA binding activity (Fox et al., 1992). Since neither fraction is yet pure, more factors may remain to be discovered. However, the poly(A) polymerase may be the only necessary component in this fraction, since it can be replaced by a pure enzyme prepared from calf thymus (Fox et al., 1992). In the case of the RNA binding activity, it is not yet clear whether distinct factors recognize AAUAAA and UUUUUAU. Indeed, it has not formally been demonstrated that the RNA binding activity is itself necessary for the reaction, although its sequence specificity very strongly suggests that this is the case.

The observation that specific proteins can be cross-linked by UV light to RNAs carrying the U-rich elements and AAUAAA suggests that those proteins play a role in polyadenylation. Proteins of 58 kD (Paris et al., 1991) and 82 kD (McGrew and Richter, 1990) have been implicated in this way. Because each crosslinks to one mRNA and not another, it is possible that polyadenylation of different mRNAs requires different positive acting factors.

## The enzymes involved in poly (A) removal: yeast PAN and its possible cousins

The enzymes that remove poly(A) in frog egg extracts have not yet been purified, although their activity is readily demonstrable (Varnum et al., in press; K. Takayama and M. Wickens, unpub. exps). At least two oocyte factors appear to be involved in deadenylation during maturation, since both nuclear and cytoplasmic components are required (Varnum et al., in press). However, these need not be direct participants in the reaction; one might simply activate the latent nuclease activity of the other.

In yeast, however, an enzyme that specifically removes poly(A) has been purified to near homogeneity (Sachs and Deardorff, 1992; Lowell et al, 1992). This enzyme, called PAN (poly(A) nuclease), requires poly(A)-binding protein (PAB) to act -- it is inactive on naked poly(A), but gains activity when poly(A) is mixed with pure PAB (Lowell et al, 1992). It appears to be a 3' to 5' exonuclease that liberates 5' mononucleotides, and is inhibited by a long non-adenosine stretch after the poly(A) tail (Lowell et al, 1992). In these respects, it seems to resemble the less well characterized deadenylation activities found in HeLa extracts (Astrom et al., 1991) and in frog oocyte during maturation (Varnum et al., in press). It seems an attractive possibility that the enzymes involved in cytoplasmic deadenylation in these very divergent organisms may be closely related, though no decisive evidence on this point yet exists.

**Relevance of the oocyte studies: the correlation between mRNA adenylation and translation is a common but not universal theme**

In addition to the studies in frog oocytes and other developing systems several other experimental approaches have demonstrated a role for poly(A) in the regulation of translation. For example, the translatability of synthetic mRNAs, differing solely in 5'-cap and/or 3'-poly(A) structures, has been compared in rabbit reticulocyte extracts and in electroporated plant, animal, and yeast cells (Munroe and Jacobson, 1990b; Gallie, 1991). The results of these experiments confirm that poly(A)$^-$ mRNAs have a reduced translational capacity and that the effect of poly(A) on translation is directly related to its length. In electroporated cells, translation is stimulated over an order of magnitude by a poly(A) tail and such stimulation is dependent on the presence of a 5'-cap on the same mRNA (Gallie, 1991). In reticulocyte extracts the extent of poly(A)-mediated stimulation is more modest (approx. twofold) and such stimulation is not cap dependent (Munroe and Jacobson, 1990b).

A direct comparison of the polysomes formed in reticulocyte extracts by poly(A)$^+$ and poly(A)$^-$ mRNAs revealed that the polysomes formed by poly(A)$^-$ mRNAs contain fewer ribosomes than those formed by poly(A)$^+$ mRNAs. These data, together with the finding that translational elongation rates along adenylated and unadenylated mRNAs are identical, establish that mRNA poly(A) tails function in translational initiation (Munroe and Jacobson, 1990b). Other experiments showed that the defect in poly(A)⁻ mRNAs is distinct from that associated with cap-deficient mRNAs and results in a reduced ability to join 60S ribosomal subunits to 48S preinitiation complexes during the later stage(s) of translational initiation (Munroe and Jacobson, 1990b).

The relationship between poly(A) tail length and translational efficiency is not absolute. There are examples of translationally inactive polyadenylated mRNAs (Raff, 1980; Rosenthal et al., 1983) as well as examples of mRNAs which appear to lose their poly(A) tracts as they become translationally active (Hruby and Roberts, 1977; Iatrou and Dixon, 1977; Kleene, 1989). Perhaps the most striking inconsistency with the correlation between adenylation status and translation has been observed in a yeast mutant with a temperature-sensitive lesion in poly(A) polymerase. After a shift to the non-permissive temperature, cells of this mutant (*pap1-1*) rapidly accumulate mRNAs lacking poly(A) tails, but overall cellular translation rates are virtually unaffected for at least two hours (Patel and Butler, 1992). These results suggest that a poly(A) tail may provide an mRNA with a competitive advantage in translation, but that, in the absence of competition, the tail is not essential.

## What mediates the translational function of poly(A)?

The translational effect of poly(A) is almost certainly mediated by the cytoplasmic poly(A)-binding protein (PAB). This conclusion is supported principally by experiments in yeast analyzing mutants of PAB or suppressors of those mutants. *In vivo* depletion of PAB by promoter inactivation or by the use of a PAB temperature-sensitive mutation results in a marked decrease in the amount of polysomes along with a concomitant increase in free ribosomal subunits, a phenotype characteristic of a translational initiation defect (Sachs and Davis, 1989). Seven independent, extragenic revertants of the ts mutation which allow translational initiation in the absence of PAB have been isolated and all of these revertants affect cellular levels of the 60S ribosomal subunit. Two revertants (*spb2-1* and *spb4-1*) have been localized, respectively, to the gene for the 60S ribosomal protein L46 and to a gene encoding a putative rRNA helicase involved in the maturation of 25S rRNA (Sachs and Davis, 1989; 1990). These data indicate that translation in the absence of functional PAB may require an alteration in the structure of the 60S ribosomal subunit, a conclusion in concordance with the function ascribed to poly(A) in reticulocyte extracts (see above).

## Poly (A) removal and mRNA instability: cause-and-effect?

Studies of the turnover of individual mRNAs in a variety of cells have indicated that at least one pathway of mRNA degradation requires that poly(A) tail removal, or extensive poly(A) shortening, precede decay of the remainder of the mRNA. For example, in somatic mammalian cells a number of short-lived mRNAs, such as the c-*fos* and c-*myc* proto-oncogene transcripts and interferon-β mRNA, are swiftly and virtually completely deadenylated as a first step in their rapid decay (Wilson and Treisman, 1988; Laird-Offringa, 1990; Shyu et al., 1991; Peppel and Baglioni, 1991). Such mRNAs typically survive in the cytoplasm for only 30-60 minutes before being degraded. In contrast, more stable mRNAs such as β-globin mRNA undergo slow and incomplete poly(A) shortening and can persist in the cytoplasm for many hours (Shyu et al., 1991). It is important to emphasize that not all pathways for rapid mRNA decay in mammalian cells involve poly(A) removal as an initial step. For example, a mutation in β-globin mRNA that causes premature translation termination markedly accelerates the turnover of this message without causing any change in its slow rate of deadenylation (Shyu et al., 1991).

The RNA elements that direct accelerated deadenylation of the c-*fos* message have been mapped to two domains, and they turn out to be the same elements that target this mRNA for rapid degradation (Wilson and Treisman, 1988; Shyu et al., 1991). One of these domains is a 75-nucleotide AU-rich sequence in the c-*fos* 3' untranslated region. This AU-rich element resembles similar elements present in a variety of labile proto-oncogene and cytokine mRNAs, such as c-*myc*, GM-CSF, and interferon-β mRNA (Shaw and Kamen, 1986). The other destabilizing domain comprises the protein-coding region of the c-*fos* transcript. Each of these c-*fos* domains alone is sufficient to direct rapid deadenylation and decay when inserted into an otherwise long-lived mRNA (Shyu et al., 1991; Schiavi et al., 1994).

The features of AU-rich destabilizing elements that are responsible for their activity are not yet clear. Most include tandemly repeated copies of the tetranucleotide AUUU, suggesting that this repeated motif and related sequences may be recognized by a cellular factor that mediates deadenylation (Shaw and Kamen, 1986). The elements in protein-coding regions that promote rapid deadenylation and decay are more poorly understood. In the case of the c-*fos* coding region, it has been shown by translational frameshifting that the destabilizing signal comprises some feature of the sequence or structure of this c-*fos* RNA segment rather than a feature of either the c-*fos* codon sequence or the nascent Fos polypeptide encoded by this segment (Wellington et al., 1993).

Given the ability of poly(A) tails to stimulate translation initiation, it is of special interest that the elements which promote deadenylation as a first step in rapid mRNA decay can be activated by translation. Thus, cis-acting mutations that inhibit translation initiation have been found to impair the ability of the c-*fos* coding region and the GM-CSF AU-rich element to mediate mRNA decay (Savant-Bhonsale and Cleveland, 1992; Aharon and Schneider, 1993; Schiavi et al., 1994). In the case of the c-*fos* coding region, it has been shown that inhibiting translation in this manner markedly slows deadenylation as well (Schiavi et al., 1994). These findings suggest a model in which the poly(A) tail, upon binding to the large ribosomal subunit during translation initiation, might be brought into close proximity to a hypothetical ribosome-associated nuclease that can be activated by sequences within the coding region or 3' untranslated region of the message.

Evidence that mRNA decay is dependent on prior poly(A) shortening has also been obtained from several independent experimental approaches in yeast. Using a transcriptional pulse procedure, comparable in principle to that used to study the mammalian c-*fos* mRNAs, Decker and Parker (1993) showed that extensive poly(A) shortening precedes decay of three different mRNAs. For one of these mRNAs (*MFA2* mRNA), 3'-UTR mutations were isolated that both reduced the rate of poly(A) shortening and promoted mRNA stabilization (Muhlrad and Parker, 1992). Other studies showed that mRNA decay intermediates that accumulated because of either a lesion in the *XRN1* (5' to 3') exonuclease or specific oligo(G) insertions all contained oligo(A) termini, i.e., no decay intermediates were found to contain long poly(A) tails (Hsu and Stevens, 1993; Decker and Parker, 1993).

A requirement for deadenylation in the decay of specific mRNAs raises the question of which enzymes are responsible for poly(A) shortening and mRNA decay and what are their relationships with each other and with PAB. Whether the PAN exonuclease (see above), known to require PAB for activity (Lowell et al, 1992), is the enzyme responsible for the poly(A) shortening steps that promote mRNA decay remains to be established. The accumulation of long poly(A) tails in the yeast *pan1-3* conditional mutant is suggestive of a role for this enzyme. Of particular interest, however, is the observation that this mutant is defective in translational initiation (Sachs and Deardorff, 1992). This fascinating twist suggests the existence of a protein complex responsible for mediating all of the effects of the poly(A) tail.

## Acknowledgments

Research in the authors' laboratories was sponsored by grants (GM31892 to MW; GM42720 to JGB; and GM27757 to AJ) from the N.I.H., by a Research Career Development Award (GM00521) to MW from the N.I.H., and by a Faculty Research Award (FRA-419) to JGB from the American Cancer Society.

## Literature Cited

**Aharon, T., and Schneider, R. J.** 1993. Selective destabilization of short-lived mRNAs with the granulocyte-macrophage colony-stimulating factor AU-rich 3' noncoding region is mediated by a cotranslational mechanism. Mol. Cell. Biol. **13**: 1971-1980.

**Astrom, J., Astrom, A. and Virtanen, A.** 1991. *In vitro* deadenylation of mammalian mRNA by a HeLa cell 3' exonuclease. EMBO J. **10**: 3067-3071.

**Bouvet, P., Paris, J., Philippe and Osborne, H.B.** 1991. Degradation of a developmentally regulated mRNA in *Xenopus* embryos is controlled by the 3' region and requires the translation of another maternal mRNA. Mol. Cell Biol. **11**: 3115-3124.

**Decker, C.J. and Parker, R.** 1993. A turnover pathway for both stable and unstable mRNAs in yeast: evidence for a requirement for deadenylation. Genes Dev. **7**: 1632-1643.

**Fox, C.A., Sheets, M.D., and Wickens, M.** 1989. Poly(A) addition during maturation of frog oocytes: distinct nuclear and cytoplasmic activities and regulation by the sequence UUUUUAU. Genes Dev. **3**: 2151-2162.

**Fox, C.A., Sheets, M.D., Wahle, E. and Wickens, M.** 1992. Polyadenylation of maternal mRNA during oocyte maturation: poly(A) addition *in vitro* requires a regulated RNA binding activity and a poly (A) polymerase. EMBO J. **11**: 5021-5032.

**Fox, C.A. and Wickens, M.P.** 1990. Poly (A) removal during oocyte maturation: a default reaction selectively prevented by specific sequences in the 3'-UTR of certain maternal mRNAs. Genes Dev. **4**: 2287-2298.

**Gallie, D.R.** 1991. The cap and poly(A) tail function synergistically to regulate mRNA translational efficiency. Genes Dev. **5**: 2108-2116.

**Hsu, C.L. and Stevens, A.** 1993. Yeast cells lacking 5'--->3' exoribonuclease 1 contain mRNA species that are poly(A) deficient and partially lack the 5' cap structure. Mol. Cell. Biol. **13**: 4826-4835.

**Huarte, J., Stutz, A., O'Connell, M.L., Gubler, P., Belin, D., Darrow, A.L. Strickland, S. and Vassalli, J.D.** 1992. Transient translational silencing by reversible mRNA deadenylation. Cell **69**: 1021-1030.

**Hruby, D.E. and Roberts, W.K.** 1977. Encephalomyocarditis virus RNA II. Polyadenylic acid requirement for efficient translation. J. Virol. **23**: 338-344.

**Iatrou, K., and Dixon, G.H.** 1977. The distribution of poly(A)$^+$ and poly(A)$^-$ protamine messenger RNA sequences in the developing trout testis. Cell **10**: 433-441.

**Kleene, K.** 1989. Poly(A) shortening accompanies the activation of translation of five mRNAs during spermiogenesis in the mouse. Development **106**: 367-373.

**Laird-Offringa, I. A., De Wit, C. L., Elfferich, P., & van der Eb, A. J.** 1990. Poly(A) tail shortening is the translation-dependent step in c-*myc* mRNA degradation. Mol. Cell. Biol. **10**: 6132-6140.

**Lowell, J.E., Rudner, D.Z., and Sachs, A.B.** 1992. 3'-UTR-dependent deadenylation by the yeast poly(A) nuclease. Genes Dev. **6**: 2088-2099.

**McGrew, L., Dworkin-Rastl, E., Dworkin, M.B. and Richter, J.D.** 1989. Poly(A) elongation during *Xenopus* oocyte maturation is required for translational recruitment and is mediated by a short sequence element. Genes Dev. **3**: 803-815.

**McGrew, L. L. and Richter, J.D.** 1990. Translational control by cytoplasmic polyadenylation during Xenopus oocyte maturation: characterization of cis and trans elements and regulation by cyclin/MPF. EMBO J. **9**: 3743-3751.

**Muhlrad, D. and Parker, R.** 1992. Mutations affecting stability and deadenylation of the yeast *MFA2* transcript. Genes Dev. **6**: 2100-2111.

**Munroe, D. and Jacobson, A.** 1990a. Tales of poly(A)--a review. Gene **91**: 151-158.

**Munroe, D. and Jacobson, A.** 1990B. mRNA poly(A) tail, a 3' enhancer of translational initiation. Mol. Cell. Biol. **10**: 3441-3455.

**Paris, J. and Richter, J.D.** 1990. Maturation-specific polyadenylation and translational control: diversity of cytoplasmic polyadenylation elements, influence of poly(A) tail size, and formation of stable polyadenylation complexes. Mol. Cell Biol. **10**: 5634-5645.

**Paris, J., Swenson, K., Piwicna-Worms, H., and Richter, J.D.** 1991. Maturation-specific polyadenylation: *in vitro* activation by p34$^{cdc2}$ and phosphorylation of a 58 kD CPE-binding protein. Genes Dev. **5**: 1697-1708.

**Patel, D. and Butler, J.S.** 1992. Conditional defect in mRNA 3' end processing caused by a mutation in the gene for poly(A) polymerase. Mol. Cell. Biol. **12**: 3297-3304.

**Peppel, K. and Baglioni, C.** 1991. Deadenylation and turnover of interferon-β mRNA. J. Biol. Chem. **266**: 6663-6666.

**Raff, R.A.** 1980. In Cell Biology: A Comprehensive Treatise, Vol. 4, Gene expression: Translation and the behavior of proteins (eds: D.M. Prescott and L. Goldstein), Academic press, NY, pp. 107-136.

**Rosenthal, E.T., Tansey, T.R., and Ruderman, J.V.** 1983. Sequence-specific adenylations and deadenylations accompany changes in the translation of maternal messenger RNA after fertilization of Spisula oocytes. J. Mol. Biol. **166**: 309-327.

**Sachs, A.B., and Davis, R.W.** 1989. The poly(A)-binding protein is required for poly(A) shortening and 60S ribosomal subunit dependent translation initiation. Cell **58**: 857-867.

**Sachs, A.B. and Davis, R.W.** 1990. Translational initiation and ribosomal biogenesis: involvement of a putative rRNA helicase and RPL46. Science **247**: 1077-1079.

**Sachs, A.B. and Deardorff, J.A.** 1992. Translation initiation requires the PAB-dependent poly(A) ribonuclease in yeast. Cell **70**: 961-973.

**Savant-Bhonsale, S. and Cleveland, D.W.** 1992. Evidence for instability of mRNAs containing AUUUA motifs mediated through translation-dependent assembly of a >20S degradation complex. Genes Dev. **6**: 1927-1939.

**Schiavi, S. C., Wellington, C. L., Shyu, A.-B., Chen, C.-Y. A., Greenberg, M. E., and Belasco, J. G.** 1994. Multiple elements in the c-*fos* protein-coding region facilitate mRNA deadenylation and decay by a mechanism coupled to translation. J. Biol. Chem. (in press).

**Shaw, G. and Kamen, R.** 1986. A conserved AU sequence from the 3' untranslated region of GM-CSF mRNA mediates selective mRNA degradation. Cell **46**: 659-667.

**Shyu, A.-B., Belasco, J. G., and Greenberg, M. E.** 1991. Two distinct destabilizing elements in the c-*fos* message trigger deadenylation as a first step in rapid mRNA decay. Genes Dev. **5**: 221-231.

**Varnum, S.M. and Wormington, W.M.** 1990. Deadenylation of maternal mRNAs during *Xenopus* oocyte maturation does not require specific cis-sequences: a default mechanism for translational control. Genes Dev. **4**: 2278-2286.

**Varnum, S.M., Hurney, C.A. and Wormington, M.** Maturation-specific deadenylation in *Xenopus* oocytes requires nuclear and cytoplasmic factors. Dev. Biol. (in press).

**Wahle, E. and Keller, W.** The biochemistry of 3'-end cleavage and polyadenylation of messenger RNA precursors. Ann. Rev. Biochem. (in press).

**Wellington, C. L., Greenberg, M. E., and Belasco, J. G.** 1993. The destabilizing elements in the coding region of c-*fos* mRNA are recognized as RNA. Mol. Cell. Biol. **13**: 5034-5042.

**Wickens, M.P.** 1990. How the messenger got its tail: addition of poly(A) in the nucleus. TIBS **15**: 277-280.

**Wickens, M.** 1992. Forward, backward, how much, when: mechanisms of poly (A) addition and removal and their role in early development. Sem. Dev. Biol. **3**: 399-412.

**Wilson, T. and Treisman, R.** 1988. Removal of poly(A) and consequent degradation of c-*fos* mRNA facilitated by 3' AU-rich sequences. Nature **336**: 396-399.

# MECHANISM AND REGULATION OF TRANSLATION INITIATION IN PROKARYOTES*

Cynthia L. Pon and Claudio O. Gualerzi
Laboratory of Genetics
Department of Biology
University of Camerino
62032 Camerino (MC), Italy

In this chapter we shall present a short summary of the most important concepts concerning the basic mechanism of translation initiation which are relevant for understanding the various types of regulation occurring at this particular stage of gene expression (for more extensive reviews see Gualerzi et al. 1990; Hartz et al. 1990; Gualerzi and Pon 1990; Gualerzi and Pon 1993). In addition, we shall try to summarize and/or update the information on regulation of translation initiation which has been presented in recent reviews (de Smit and van Duin 1990; Gold 1988; McCarthy and Gualerzi 1990).

## The basic mechanism of translation initiation

In most, if not all cases, initiation starts with the small (30S) ribosomal subunit interacting, in random order, with the Translation Initiation Region (TIR) of the mRNA (see below) and with the initiator fMet-tRNA$^{fMet}$. The two ribosomal ligands are initially bound to separate and independent sites (i.e. they do not interact with each other) until a first-order, rate-determining step, presumably consisting of (or accompanied by) a conformational transition of the 30S ribosomal subunit, causes the two ligands to interact (codon-anticodon interaction) "locking" them in a more or less stable "30S initiation complex". Binding of a 50S ribosomal subunit to this complex yields a "70S initiation complex" in which fMet-tRNA eventually occupies the ribosomal P site and the A site is ready to accept the EFTu-GTP-aminoacyl tRNA complex specified by the second codon of the mRNA. A scheme illustrating the initiation pathway is presented in Fig 1.

## The role of the initiation factors

In the process described above, the 30S subunit is assisted by three fairly well characterized

NATO ASI Series, Vol. H 97
Post-transcriptional Control of Gene Expression
Edited by Orna Resnekov and Alexander von Gabain
© Springer-Verlag Berlin Heidelberg 1996

proteins, the translation Initiation Factors IF1, IF2 and IF3. Each of these factors is bound in single copy to a specific site of the 30S subunit and plays a specific and important (all three IFs are essential for cell survival) role in controlling the kinetic parameters of the aforementioned rate-limiting step.

In contrast to what was initially believed by several investigators, IF3 (20.5 kDa) does not play a direct role in binding the mRNA to the 30S subunit. Instead, IF3 acts by accelerating both formation and dissociation of the 30S initiation complex (step C in Fig. 1); IF3 stimulates more or less to the same extent the binding of any aminoacyl-tRNA (e.g. Phe-tRNA, Val-tRNA, fMet-tRNA) to the 30S subunit in response to the appropriate templates, but can differentiate among the various possible complexes initially formed and dissociate much faster those which are non-canonical. Thus, IF3 stimulates protein synthesis if the correct ribosomal ligands (mRNA TIR and fMet-tRNA) are present in the complex, but dissociates rapidly complexes accidentally formed with other aminoacyl-tRNAs at non-initiation sites of the mRNA, thereby acting as a fidelity factor to prevent any spurious complex from becoming "70S initiation complex" and entering the elongation pathway. In addition to its role in 30S initiation complex formation, IF3 shifts to the right the equilibrium 70S $\rightleftharpoons$ 30S + 50S by binding to the 30S subunits, thus providing a continuous supply of free 30S subunits to fuel the initiation process.

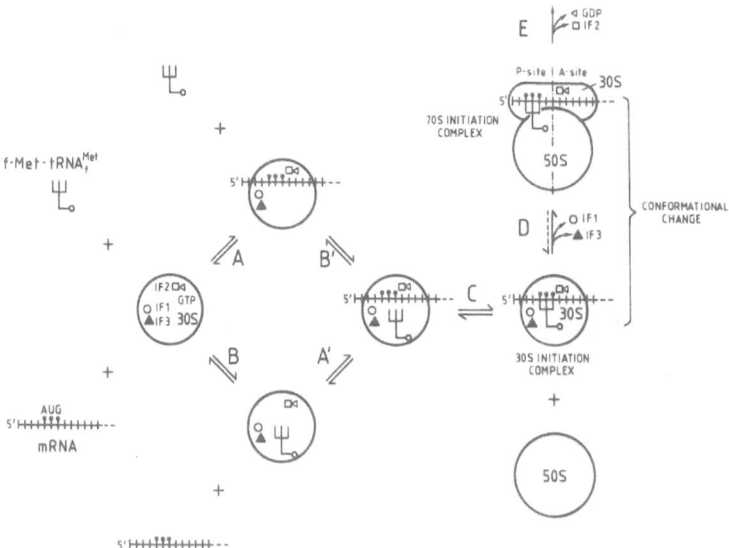

Figure 1 Translation initiation pathway in prokaryotes.

IF2 (~97 kDa) binds to the 30S subunits in the presence of a GTP molecule, and its role is somewhat complementary to that of IF3. In fact, IF2 stimulates the association of the two ribosomal subunits (step D in Fig. 1) and slows down the dissociation of all aminoacyl-tRNAs from their coded complexes with the 30S subunit (step C in Fig. 1). Yet this factor also favors the formation of genuine initiation complexes by accelerating the binding of fMet-tRNA (step C and, probably, B and B' in Fig. 1) much more than that of aminoacyl-tRNAs not having a blocked $\alpha$ $NH_2$ group. Upon formation of the 70S initiation complex, IF2 promotes the adjustment of the fMet-tRNA molecule from a "pre-P site" to the bona fide "P site" (step E in Fig. 1). This activity is kinetically stimulated by the presence of GTP and, presumably, by its IF2-dependent hydrolysis (La Teana et al. manuscript in preparation). Finally, it should be mentioned that the available evidence suggests that IF2 also plays a major role in adjusting the mRNA on the 30S subunit shifting it from a "stand-by" site where it is originally bound to another site where codon-anticodon interaction is presumably favored (Gualerzi and Pon 1993). Thus, the properties of IF2 are somewhat reminiscent of those of the two elongation factors EF-Tu and EF-G combined.

The smallest factor, IF1 (8 kDa), stimulates the activity of both IF2 and IF3 in the formation of the 30S initiation complex (step C in Fig. 1). Furthermore, IF1 has a profound influence on the $K_{ass}$ of IF2-30S complex; its presence ensures the very stable interaction required in the formation of 30S initiation complex, while its ejection, triggered by the binding of the 50S subunit (step D in Fig. 1), leaves IF2 in a metastable binding site on the 70S ribosomes, from which it is ejected upon GTP hydrolysis. Overall the most important function of IF1 may well be this modulation of the IF2 affinity for the ribosome.


Characteristics of the mRNA translation initiation region (TIR)


To start protein synthesis, the 30S ribosomal subunit must bind to a specific region of the mRNA; it is indeed possible to demonstrate that the initiating ribosome protects a segment of mRNA spanning from about -20 to +13 (Hartz et al. 1988). This sequence, often called ribosomal binding site (RBS), may or may not be sufficient to initiate translation; there are cases of mRNAs containing strong but translationally inactive RBS, while there are many known cases of nucleotides, upstream or downstream of the RBS, which greatly influence (either positively or negatively) the rate of initiation of particular mRNAs. Thus, rather than "RBS", we prefer to use the term "translation initiation region" (TIR), which has a broader functional definition. It should be obvious, however, that to be active a TIR must also contain some kind of RBS.

The distinguishing features of the mRNA TIRs have been identified primarily from comparison and statistical analyses of the sequences of hundreds of genuine translational starts (Scherer et al. 1980; Schneider et al. 1986; Schneider and Stephens 1990; Rudd and Schneider 1992). This approach revealed the existence of several elements displaying nonrandom primary structures within the RBS of the mRNA TIRs. The most important of these elements are the initiation triplet and, separated from it by a spacer of variable length, the Shine-Dalgarno (SD) sequence (typically UAAGGAGG) which is complementary to the anti-SD sequence found at the 3' end of 16S rRNA.

Several studies have addressed the problem of defining, in both absolute and relative terms, the roles played by the individual TIR elements in determining overall translational efficiency of various types of mRNAs (Hartz et al. 1991; Ringquist et al. 1992; Vellanoweth and Rabinowitz 1992).

After recalling that it is the interplay between two or more of the different elements present which determines the quality of each TIR, we shall briefly and separately describe these elements and try to assess their relevance for translational efficiency and/or control.

## Initiation codon

The most recent compilation of mRNA TIRs derived from 1055 genes of *Escherichia coli*, accounting for approximately 38% of the entire genome (Rudd and Schneider 1992), shows the use of four possible initiation triplets in the following proportions: AUG, 91.94%; GUG, 6.73%; UUG, 1.23% and AUU, found only once in the *infC* (IF3) gene. The reason for this degeneracy of the initiation triplet is not clear. In three separate studies (Ringquist et al. 1992; Vellanoweth and Rabinowitz 1992; Sussman and Simons 1993), it was shown that the rare codons at the beginning of reporter genes are less efficient than AUG (from just a few fold to over two orders of magnitude). Nevertheless, the fact that the list of genes beginning with the rare codons includes some of those most highly expressed suggests that the rare codons are not used to maintain a low level of translation. Substitution of the rare codons of some genes with the canonical AUG triplet was found to affect the in vivo expression in different ways ranging from a large (>30-fold) stimulation following the AUU → AUG change in *infC* (Brombach and Pon 1987) to a substantial (7-fold) inhibition following the GUG → AUG change in *rpsM* (Spedding et al. 1993).

At least in one case, namely with the AUU start triplet of *infC*, the rare initiation codon represents a regulatory signal for the translational autorepression by IF3 (Gold et al. 1984, Butler et al. 1987, Hartz et al. 1989, La Teana et al. 1993). Compared to AUG, the AUU triplet does not limit per se mRNA translation in vitro, but renders translation initiation susceptible to the presence of excess IF3, especially in the presence of other competing mRNAs. The inhibition by IF3 occurs via the increased dissociation of 30S initiation complexes containing the rare triplet (step

C of Fig. 1), presumably coupled to interference by excess IF3 with the association of 50S subunits with the 30S initiation complex (step D in Fig. 1) (La Teana et al. 1993). A recent study showed that IF3 can indeed discriminate between different potential initiation triplets in vivo: an IF3 mutant with an amino acid replacement at Tyr75 allowed higher levels of expression of a reporter gene beginning with unusual initiation codons such as GUG, AUU, AUG, AUA and ACG. On the other hand, expression from other initiation triplets was either not affected by IF3 (AUG, GUG, UUG) or extraordinarily low under all circumstances (Sussman and Simons 1993).

The Shine-Dalgarno sequence

Basepairing in vivo between the mRNA SD sequence and anti-SD sequence of 16S rRNA has been demonstrated by showing that ribosomal mutants containing nucleotide changes in the 1535-1540 region of 16S rRNA can compensate for translational inefficiency resulting from defects in the SD sequence of the template (Hui and de Boer 1987; Jacob et al. 1987). The importance of the SD sequence and of its accessibility for the 30S subunits in determining the extent to which individual mRNAs are translated in vivo is further documented by a vast literature; also recent studies aimed at a more systematic and empirical definition of the TIR determinants of translation efficiency have confirmed its importance (Hartz et al. 1991; Ringquist et al. 1992; Vellanoweth and Rabinowitz 1992). Nevertheless, the SD interaction appears to be neither mechanistically important for translation initiation in particular, nor essential for translation in general, since mRNAs with and without SD sequence are translated in vitro at the same rate, with the same accuracy and with the same dependence on initiation factors, provided that the mRNA concentration is properly adjusted (Calogero et al. 1988). In agreement with these data are the findings that 30S subunits lacking the anti-SD sequence are able to initiate natural mRNA translation at the correct start sites (Melançon et al. 1990b; Afonina et al. 1991). That the influence of the SD interaction on translation is greater in vivo than in vitro is indicated by the fact that the same mutation in the SD sequence of phage T7 0.3 mRNA which produces a drastic reduction of translation in vivo (Dunn et al. 1978) has only a marginal effect in vitro (Ohsawa et al. 1984). Thus, it has been proposed that the function of the SD sequence is simply that of increasing the local concentration of the initiation triplet near the decoding site of the ribosome (Calogero et al. 1988). The notion that the SD sequence is not essential per se in translation is further strengthened by the existence of mRNAs with very weak (i.e. not better than any Gly codon) or completely lacking any sequence 5' of the initiation triplet. Among the latter type of mRNAs there is phage λ cI mRNA (when transcribed from the $P_{RM}$ promoter), the tetR mRNA of transposon Tn1721 (Baumeister et al. 1991) and, despite the claim that a strong SD sequence is essential for translation in gram positive bacteria (Vellanoweth and Rabinowitz 1992), some well expressed mRNAs of *Streptomyces* (Jones et al. 1992 and references therein). In conclusion, since a TIR having an extended SD sequence can be expected to compete better with other TIR and non-TIR sequences for 30S binding, it is obvious that availability/unavailability and good/poor quality of the SD sequence may play an important role in determining the efficiency and/or the regulation of translation initiation. These parameters are expected to affect the position of the A and A' equilibria in the pathway shown in Fig. 1 and it is certainly not by chance that the majority of the known specific translational regulatory loops seem to influence, either directly or indirectly, this step of initiation (see below). Generalized regulation of translation initiation, on the other hand, could be expected to occur through the control of the B and B' equilibria (Fig. 1) if global responses such as those ensuing nutritional shiftdown or cold shock result in oscillations of the intracellular levels of fMet-tRNA. The regulation of the expression of cI gene can be regarded as an extreme example of a control mechanism which depends on the availability of the SD sequence: transcription from $P_{RE}$ or from $P_{RM}$ produces two mRNAs with and without SD sequence expressing the high and low levels of repressor required for establishing and maintaining lysogeny, respectively.

## The spacer

In natural mRNAs the spacing between the SD sequence and the initiation triplet varies from 5 to 13 bases (Rudd and Schneider 1992). It is normally accepted that efficient translation initiation occurs with mRNAs having little or no secondary structure in the whole TIR. Thus, it is not surprising that even if no specific sequence seems to be preferred in the spacer, in this region there is nonetheless a bias in favor of the less "structurogenic" nucleotides such as A and U. Systematic studies have demonstrated that translational efficiency in both *E. coli* and *B. subtilis* (Ringquist et al. 1992; Vellanoweth and Rabinowitz 1992) is sensitive to spacer length with a rather broad optimum (between 6-9 nts) which depends, to some extent, upon the nature of the SD sequence. Even long spacers (i.e. 11-13 nts) give rise to fairly high expression but no translation occurs if the spacer is too short (i.e. <4); furthermore, even though *E. coli* and *B. subtilis* ribosomes exhibited similar spacer-length optima, the ribosomes from the gram positive organism seemed to translate better mRNAs with shorter spacers.

## Additional TIR elements

As mentioned above, in addition to the SD sequence and the initiation triplet, several other regions of the mRNA TIR, both upstream and downstream of the initiation triplet, appear to have sequences which are significantly non-random. Even though the statistical relevance of these extra elements seemed to fade away when the bacteriophage mRNA sequences were omitted from the analysis (Rudd and Schneider 1992), it is generally expected that other TIR recognition signals might be used by some subsets of mRNAs. The existence of mRNAs having weak SD elements or completely lacking them emphasizes the likelihood that other regions of the mRNA TIR may serve as alternative recognition elements to direct the binding of ribosomes to the mRNA initiation site and to determine translation efficiency. Indeed, many sequences have been proposed to play this role. Among these, the one with the most compelling evidence is the so called "downstream box" which corresponds to a sequence between +15 and +26 downstream of the initiator codon of phage T7 0.3 mRNA. This "SFF box" (from the initials of the original proposers) is also present in other highly expressed genes and was found to act as a translational enhancer when transferred to other mRNAs (Sprengart et al. 1990). A similar SFF box is also found downstream the initiation triplet of cI mRNA and may well account for the interaction of this mRNA with the ribosome in the absence of the SD - anti-SD interaction. SFF boxes have also been found in a large number of aminoacyl-tRNA synthetase genes of *E. coli* and to be responsible for the high level of expression of at least two of them (i.e. *glnS* and *lysU*) (Faxen et al. 1991; Ito et al. 1993). Finally, the SFF box is the target of a regulatory loop controlling translation. Thus, a segment of the coding region of *rpoH* mRNA (the heat-shock transcriptional factor $\sigma^{32}$) functions as a cis-acting antisense RNA forming a secondary structure with a cis-acting translational enhancer of the same mRNA. The enhancer was identified as the SFF box and formation of the secondary structure was found to be the switch which allows the transient expression of *rpoH* during heat-shock response (Kamath-Loeb and Gross, 1991; Nagai et al. 1991).

An additional case of translational regulation which may have its basis in another TIR element concerns the translational initiation lag experienced by some mRNAs during heat shock. This potentially interesting phenomenon seems to depend upon the presence, in the responsive mRNAs, of the rather loose consensus sequence $GAGGAA(N)_{3-5}AUG$ partially overlapping the initiation triplet and SD sequence (Kuriki 1990).

Another potential mRNA-ribosome interaction could occur between nucleotides 1-18 of 16S rRNA and at least three nucleotides of an A/U-rich region present between +4 and +25 of some mRNAs (Petersen et al. 1988). The finding that deletions in the 5' region of 16S rRNA results in 30S subunits with a reduced capacity to bind mRNA (Melançon et al. 1990a) suggests that this interaction may indeed occur.

The list of the TIR elements which may play a role in translation initiation also includes short runs of U's found upstream of the SD sequences of several mRNAs; it has been suggested that an efficient interaction of these mRNAs with the 30S subunit is due to these sequences which interact preferentially with r-protein S1 (Boni et al. 1991, Hartz et al. 1991).

Finally, at the end of this section it seems relevant to mention that, depending on its particular context, the nature of the first codon following the initiation triplet may also influence the level of translation; when the initiation codon was GUG or UUG, either AAA or GCU in the second position was found to improve the level of translation (Ringquist et al. 1992).

Trans acting translational repressors

Both RNA and proteins can act as diffusible trans acting inhibitors of translation initiation on specific target mRNAs. As mentioned above, in most of the cases (but there are very interesting exceptions) these inhibitors sequester the TIR in an inactive conformation, or compete out the 30S subunit from its mRNA binding site and thereby interfere with initiation shifting to the left equilibria A and A' (Fig. 1).

a) RNA-binding protein repressors.

Important examples of this class of inhibitors are those ribosomal (r-) proteins which act as repressors of their own synthesis as well as of all (or nearly all) other r-proteins encoded by the same operon. In view of the fact that even a superficial description of this vast subject would be lengthy and complex and that excellent reviews of this topic appeared not too long ago (Lindahl and Zengel 1986; Jinks-Robertson and Nomura 1987), we shall provide here only a brief description of the mechanism of translational control in two r-protein operons, namely the α and the S15 operons, for which substantial progress has been made in recent years (Tang and Draper 1989; Tang and Draper 1990; Spedding et al. 1993; Philippe et al. 1990; Philippe et al. 1993). In the case of the α operon, which encodes four r-proteins (S13, S11, S4 and L17), translational repression is provided by S4, one of the first proteins to bind to 16S rRNA during the assembly of the 30S subunits. In the S15 operon, which includes also the *pnp* (poly-nucleotide phosphorylase) gene, translational repression is due to S15, which is also a primary 16S rRNA-binding protein. In addition to the appropriate r-protein, regulation of both operons also involves peculiar pseudoknot structures of the mRNAs which were revealed by the use of several structure-specific probes. In both cases, mutations destroying the pseudoknot structure were found to abolish the autorepression. In the α operon the target of S4 is within a 139 nucleotides-long segment of the TIR of *rpsM* (S13) mRNA (the first cistron of the operon). This segment of RNA displays a complex double-pseudoknot structure in which the first few codons of S13 are engaged in a tertiary structure involving a fairly large number of nucleotides of the 5' leader region of the mRNA. In alternative to the 16S rRNA, S4 binds to this structure and stabilizes it. In vivo and in vitro studies with *rpsM::lacZ* fusions have shown, however, that several mutations leaving unaffected the affinity of the mRNA for S4 resulted in a substantial or complete loss of translational repression by this protein. These mutations were found to disrupt the linkage between two mRNA sites (i.e. the S4 binding site and, presumably, the 30S binding site). Thus, the translational repression by S4 is explained by an allosteric mechanism coupling two mRNA sites linked by secondary and/or tertiary interactions; when the S4 repressor binds to its site on the mRNA, an inactive conformation (perhaps unable to undergo step C of Fig. 1) is induced in the ribosomal binding site (Tang and Draper 1989; Tang and Draper 1990; Spedding et al. 1993).

Particularly interesting also is the mechanism by which protein S15 regulates its own translation. In fact, it has been shown that *rpsO* mRNA exists in two alternative conformations, one involving the formation of two stem-loops and the other forming a pseudoknot. Protein S15 binds to the pseudoknot form and shifts the equilibrium towards this conformation. Unlike other translational repressors, however, S15 does not prevent the binding of the mRNA to the 30S subunit and translational repression is not due to the inability of one of the two mRNA conformers to bind to the 30S subunit. On the contrary, S15 stabilizes the pseudoknot and stimulates the binding of this mRNA conformer to the 30S subunit which, in turn, contributes a further stabilization. In the presence of S15, however, the mRNA remains trapped in a non-productive pre-ternary complex which also contains the initiator tRNA. Thus, S15 inhibits translation of its mRNA by freezing the ribosome in the otherwise unstable kinetic intermediate, while in the absence of the protein the transition pre-ternary complex → ternary complex (step C in Fig. 1) can proceed without hindrance (Philippe et al. 1990; Philippe et al. 1993).

An interesting case of molecular mimicry is responsible for the translational autorepression of *E. coli* threonyl-tRNA synthetase (ThrRS). The TIR of *thrS* mRNA contains four structural domains. Two of these domains (i.e. domains 2 and 4) are analogous to the anticodon and to the acceptor arms of tRNA[thr], respectively. In the absence of its normal substrate (i.e. tRNA[thr]), ThrRS recognizes and binds to these structures of the mRNA preventing translation, presumably by interfering with the mRNA-ribosome interaction (Brunel et al. 1992).

The sophistication of the molecular mimicry mechanism was elegantly demonstrated showing that mutations in the anticodon-like domain of the mRNA, switching the identity rules of tRNA[thr] and making it similar to tRNA[metf], renders the translation of *thrS* mRNA insensitive to ThrRS but sensitive to methionyl-tRNA synthetase repression (Graffe et al. 1992; Romby et al. 1992).

Other well known examples of this class of translational inhibitors are the coat protein of bacteriophage R17 (Bernardi and Spahr 1972; Beckett and Uhlenbeck 1988), and three gene products of bacteriophage T4, namely the DNA polymerase (Andrake et al. 1988), the SSB protein (product of gene 32) (McPheeters et al. 1988) and the RegA protein (Winter et al. 1987). While the R17 coat protein represses translation of the RNA replicase cistron, all the T4 proteins are autorepressors (i.e. they act upon their own mRNA). Furthermore, all these repressors, with the exception of RegA which acts as a translational inhibitor of several mRNAs, have a select target in a single type of mRNA.

b) antisense (mic) RNAs.

The first reports dealing with this special type of translational control appeared in the literature approximately ten years ago. They described the negative translational regulation of the IS10 transposase, responsible for multicopy inhibition of this transposable element (Simons and Kleckner 1983) and the regulation of the *ompF* (outer membrane protein) expression in *E. coli* (Mizuno et al. 1984). In both cases, regulation was due to trans acting antisense or mic (mRNA-interfering-complementary) RNAs causing the functional inactivation of their target mRNAs by base-pairing with their respective TIRs. Both micRNAs (i.e. pOUT RNA and *micF*) are approximately the same size (i.e. 180 and 174 nucleotides, respectively). The first one, resulting from transcription in the opposite orientation of the target gene, displays perfect complementarity to its target, while *micF*, transcribed from an unlinked gene mapping in a distant position of the chromosome, forms an RNA duplex containing several mismatches.

Following these early reports, other genes whose expression is controlled by micRNAs have been described. These include the *ant* mRNA of phage P22 (Liao et al. 1987; Wu et al. 1987), the *hok* mRNA controlling the segregation of plasmid R1 (Gerdes et al. 1988) and the *traJ* mRNA involved in plasmid R100 DNA transfer (Dempsey 1987). The expression of these three mRNAs is regulated by *sar* RNA, *sok* RNA and *finP* RNA, respectively.

A somewhat special case of micRNA-mediated translational control is that involving phage λ oopRNA (77 nucleotides) which is complementary to the promoter-distal portion of the phage λ cII mRNA. Like the other micRNAs, also oopRNA forms a secondary structure with the target mRNA. Translation is not inhibited because of an interference with mRNA-ribosome interaction, however, but because the formation of the RNA duplex renders the cII mRNA susceptible to RNase III digestion (Krinke and Wulff 1987; Takayama et al. 1987).

Following the discovery of natural antisense RNAs, several attempts have been made to inhibit the in vivo expression of select genes using artificial micRNAs. Even though the use of this technique has been less successful in prokaryotic than in eukaryotic systems (this is probably due to the lability of the majority of the bacterial mRNA targets), some interesting cases of control of gene expression by artificial micRNAs has also been reported in bacteria and bacteriophages (Coleman et al. 1984; Hirashima et al. 1986).

In addition to the trans acting antisense RNAs, also cis acting antisense RNAs can form mRNA secondary structures which may modulate translation initiation. Aside from the aforementioned case of the heat shock activation of *rpoH* mRNA translation, the growth-rate-dependent regulation of *gnd* (6-phosphogluconate dehydrogenase) mRNA translation represents another well-known example of this type of posttranscriptional control. In this case regulation is mediated by a segment of the mRNA coding region between codons 48 and 118) which has the properties of an "internal complementary sequence" (ICS) and can form a secondary structure with a portion of the TIR, both upstream and downstream of the initiation triplet (which is not engaged in the secondary structure, however). Duplex-destabilizing single base changes of the ICS and alterations of its position upstream of the 13th codon resulted in growth rate derepressed phenotype. It has been suggested that secondary structure formation in the *gnd* mRNA is controlled by the availability of the ICS which depends, in turn, on the rate of translation set by the cellular growth rate (Carter-Muenchau and Wolf 1989).

Translational activators

Regulation of translation is generally negative. Nonetheless, some very interesting examples of positive regulation have been described and they are outlined in this section. In bacteriophage Mu, *mom* is a non essential gene, located near *gin* in the β region of the phage genome. *Mom* encodes a specific and rare DNA-modification function which confers resistance to both Mu and host DNA vis-a-vis several restriction endonucleases. The expression of *mom* is subject to several types of control, one of them being translational activation of its mRNA by Com, a small (7 kDa) Zn-finger protein which is expressed in the late phase of Mu development. Com protein does not influence the synthesis of Mom mRNA but functions as a trans-acting mRNA binding protein. It has been suggested that Mom mRNA exists in two conformations: one (inactive) in which the GUG initiation codon is sequestered in a double-stranded RNA stem and another (active) in which the TIR is accessible to the ribosomes. Binding of Com to its target, which has been precisely mapped in the mRNA TIR, just upstream of the SD sequence, shifts the equilibrium towards the active conformation of the mRNA and allows its translation (Wulczyn et al. 1989; Hattman et al. 1991; Wulczyn and Kahmann 1991).

Regulation of expression of cII and cIII genes, whose products are involved in the establishment of bacteriophage λ lysogeny, provides two additional examples of translational activation. The product of the cIII gene (6 kDa) stabilizes the unstable cII protein which, in turn, is required for the host polymerase to transcribe the cI (repressor) gene from the otherwise inactive $P_{RE}$ promoter. The TIR of cIII mRNA contains a fairly long stretch of nucleotides, just upstream of the SD sequence, which are required for the translation of cIII::*lacZ* fusions in vivo and of cIII::bGH in vitro (the fusion with the bovine growth hormone is necessary to improve the detection of the cII protein). This essential TIR element of cIII can form a stem-loop structure with the properties of an inefficient RNase III processing site and expression of cIII in vivo is drastically reduced in mutant hosts lacking RNase III as well as in a mutant in which the RNase III recognition site has been inactivated. Cleavage at this stem-loop, however, occurs in vivo in only 10-20% of the molecules and the processed mRNA is inactive in in vitro translation. These data suggest that the nuclease binds to the upstream region of the mRNA and, without cleaving it, exposes the ribosomal binding site which is otherwise sequestered in an inactive conformation (Altuvia et al. 1987).

In the case of cII expression, the DNA-binding protein IHF (host integration factor) may act as a translational activator. IHF is required for the integration and excision of phage λ DNA by virtue of its strong capacity to introduce sharp bends in the DNA. Stimulation of cII protein synthesis by purified IHF was observed in vitro using cell-free extracts obtained from *E. coli himA* mutants. These results were substantiated by in vivo experiments with both operon and gene fusion between cII and *lacZ* or *galE* which indicated that the stimulation by IHF occurs at the translational and not at the transcriptional level. Although the exact mechanism of the activation has not yet been clarified, the target of the translational activation has been located within the TIR, upstream of the initiation triplet of cII mRNA. Failure to translate cII was also found to activate a transcriptional terminator located within this gene and to produce polar effects on the transcription downstream (Mahajna et al. 1986). In addition to this positive translational control, cII expression is also under negative translational control by oopRNA (see above).

As mentioned above, the phage λ cI repressor can be expressed either at high (during establishment of lysogeny) or at low (during maintenance of lysogeny) level depending on whether it is translated from a long mRNA transcribed from $P_{RE}$ or from a short (leaderless) mRNA transcribed from $P_{RM}$. Thus, considering that also the expression of the *int* gene is "retroregulated" at the mRNA level by RNase III cleavage, (Gottesman et al. 1982), it appears that translational control plays a central role over the entire λ lytic/lysogenic switch.

Finally, another case of translational activation has been identified in yeast mitochondria. In these organelles of prokaryotic origin the translation of the mRNA encoding coxIII (subunit III of cytochrome c oxidase) requires the interaction of a nuclear-encoded ribosomal protein (PET123 protein) with a specific translational activator protein (PET122). This activator recognizes specifically the 5' untranslated region of coxIII mRNA and, in turn, allows the small ribosomal subunit to bind and translate this mRNA (Costanzo and Fox 1988; McMullin et al. 1990).

## Acknowledgments

This work was supported by grants from the Italian CNR (PF Ingegneria Genetica) to C.O.G. and from the Italian MURST to both authors.
*In memoriam of mathematician Dr. Orlando Gualerzi, 40 years after his death.

## References

Afonina E, Chichkova N, Bogdanova S and Bogdanov A (1991) 30S ribosomal subunits with fragmented 16S rRNA: a new approach for structure and function study of ribosomes. Biochimie 73:777-787

Altuvia S, Locker-Giladi H, Koby S, Ben-Nun O and Oppenheim AB (1987) RNase III stimulates the translation of the cIII gene of bacteriophage λ. Proc Natl Acad Sci USA 84:6511-6515

Andrake M, Guild N, Hsu T, Gold L, Tuerk C and Karam J (1988) DNA polymerase of bacteriophage T4 is an autogenous translational repressor. Proc Natl Acad Sci USA 85:7942-7946

Baumeister R, Flache P, Malefors O, Gabain Av and Hillen W (1991) Lack of a 5' non coding region in Tn1721 encoded tetR mRNA is associated with a low efficiency of translation and short half-life in E. coli. Nucleic Acids Res 19:4595-4600

Beckett D and Uhlenbeck OC (1988) Ribonucleoprotein complexes of R17 coat protein and a translation operator analog. J Mol Biol 204:927-938

Bernardi A and Spahr PF (1972) Nucleotide sequence at the binding site for coat protein on RNA of bacteriophage R17. Proc Natl Acad Sci USA 69:3033-3037

Boni IV, Isaeva DM, Musychenko ML and Tzareva NV (1991) Ribosome messenger recognition: mRNA target sites for ribosomal protein S1. Nucleic Acids Res 19:155-162

Brombach M and Pon CL (1987) The unusual translation initiation codon AUU limits the expression of the infC (initiation factor IF3) gene of E. coli. Mol Gen Genet 208:94-100

Brunel C, Caillet J, Lasage P, Graffe M, Dondon J, Moine H, Romby P, Ehresmann C, Ehresmann B, Grunberg-Manago M and Springer M (1992) Domains of the E. coli threonyl-tRNA synthetase translational operator and their relation to threonine tRNA isoacceptors. J Mol Biol 227:621-634

Butler JS, Springer M and Grunberg-Manago M (1987) AUU to AUG mutation in the initiator codon of the translation initiation factor IF3 abolishes translational autocontrol of its own gene infC in vivo. Proc Natl Acad Sci USA 84:4022-4025

Calogero RA, Pon CL, Canonaco MA and Gualerzi CO (1988) Selection of the mRNA translation initiation region by E. coli ribosomes. Proc Natl Acad Sci USA 85:6427-6431

Carter-Muenchau P and Wolf Jr RE (1989) Growth-dependent regulation of 6-phosphogluconate dehydrogenase level mediated by an anti-Shine-Dalgarno sequence located within the E. coli gnd structural gene. Proc Natl Acad Sci USA 86:1138-1142

Coleman J, Green PJ and Inouye M (1984) The use of RNA complementary to specific mRNAs to regulate the expression of individual bacterial genes. Cell 37:429-436

Costanzo MC and Fox TD (1988) Specific translational activation by nuclear gene products occurs in the 5' untranslated leader of yeast mitochondrial mRNA. Proc Natl Acad Sci USA 85:2677-2681

Dempsey WB (1987) Transcript analysis of the plasmid R100 traJ and finP genes. Mol Gen Genet 209:533-544

de Smit M and van Duin J (1990) Control of prokaryotic translation initiation by mRNA secondary structure. Progr Nucl Acid Res Mol Biol 38:1-35

Dunn JJ, Buzash-Pollert E and Studier FW (1978) Mutations of bacteriophage T7 that affect initiation of synthesis of the gene 0.3 protein. Proc Natl Acad Sci USA 75:2741-2745

Faxen M, Plumbridge J and Isaksson LA (1991) Codon choice and potential complementarity between mRNA downstream of the initiation codon and bases 1471-1480 in 16S rRNA affects expression of glnS. Nucleic Acids Res 19:5247-5251

Gerdes K, Helin K, Christensen OW and Loebner-Olesen A (1988) Translational control and differential RNA decay are key elements regulating postsegregational expression of the killer protein encoded by the parB locus of plasmid R1. J Mol Biol 203:119-129

Gold L, Stormo G and Saunders R (1984) E. coli translational initiation factor IF3: a unique case of translational regulation. Proc Natl Acad Sci USA 81:7061-7065

Gold L (1988) Posttranscriptional regulatory mechanism in Escherichia coli. Annu Rev Biochem 57:199-203

Gottesman M, Oppenheim A and Court D (1982) Retroregulation: control of gene expression from sites distal to the gene. Cell 29:727-728

Graffe M, Dondon J, Caillet J, Romby P, Ehresmann C, Ehresmann B and Springer M (1992) The specificity of translational control switched with transfer RNA identity rules. Science 255:994- 999

Gualerzi CO, La Teana A, Spurio R, Canonaco MA, Severini M and Pon CL (1990) Initiation of protein biosynthesis in prokaryotes: recognition of mRNAs by ribosomes and molecular basis for the function of the initiation factors. In: Hill WE, Dahlberg A, Garrett RA, Moore PB, Schlessinger D and Warner JR (eds) The ribosome. Structure, function and evolution. Am Soc Microbiol, Washington D.C. pp 281-291

Gualerzi CO and Pon CL (1990) Initiation of mRNA translation in prokaryotes. Biochemistry 29:5881-5889

Gualerzi CO and Pon CL (1993) mRNA-ribosome interaction during initiation of protein synthesis. In: Zimmermann RA and Dahlberg AE (eds) Ribosomal RNA. CRC Press, Inc, in press

Hartz D, McPheeters DS, Traut R and Gold L (1988) Extension inhibition analysis of translation initiation complexes. Meth Enzymol 164:419-425

Hartz D, McPheeters DS and Gold L (1989) Selection of the initiator tRNA by E. coli initiation factors. Genes Dev 3:1899-1912

Hartz D, McPheeters DS and Gold L (1990) From polynucleotide to natural mRNA translation initiation: function of E. coli initiation factors. In: Hill WE, Dahlberg A, Garrett RA, Moore PB, Schlessinger D and Warner JR (eds) The ribosome. Structure, function and evolution. Am Soc Microbiol, Washington D.C. pp 275-280

Hartz D, McPheeters DS and Gold L (1991) Influence of mRNA determinants on translation initiation in E. coli. J Mol Biol 218:83-97

Hattman S, Newman L, Murthy HMK and Nagaraja V (1991) Com, the phage Mu *mom* translational activator, is a zinc-binding protein that binds specifically to its cognate mRNA. Proc Natl Acad Sci USA 88:10027-10031

Hirashima A, Sawaki S, Inokuchi Y and Inouye M (1986) Engineering of the mRNA-interfering complementary RNA immune system against viral infection. Proc Nat Acad Sci USA 83:7726-7730

Hui A and de Boer HA (1987) Specialized ribosome system: preferential translation of a single mRNA species by a subpopulation of mutated ribosomes in E. coli. Proc Natl Acad Sci USA 84:4762-4766

Ito K, Kawakami K and Nakamura Y (1993) Multiple control of E. coli lysyl-tRNA synthetase expression involves a transcriptional repressor and a translational enhancer element. Proc Natl Acad Sci USA 90:302-306

Jacob WF, Santer M and Dahlberg AE (1987) A single base change in the Shine-Dalgarno region of 16S rRNA of E. coli affects translation of many proteins. Proc Natl Acad Sci USA 84:4757-4761

Jones RL, Jaskula JC and Janssen GR (1992) In vivo translational start site selection by prokaryotic ribosomes in the absence of a Shine-Dalgarno interaction. J Bacteriol 174:475-34760

Jinks-Robertson S and Nomura M (1987) Ribosomes and tRNA. In Neidhardt FC, Ingraham JL, Brooks Low K, Magasanik B, Schaechter M and Umbarger HE (Eds) *Escherichia coli* and *Salmonella typhimurium*. Cellular and molecular biology. Am Soc Microbiol, Washington, DC pp 1358-1385

Kamath-Loeb AS and Gross CA (1991) Translational regulation of $\sigma^{32}$ synthesis: requirement for an internal control element. J Bacteriol 173:3904-3906

Krinke L and Wulff DL (1987) OOP RNA, produced from multicopy plasmids, inhibits cII gene expression through an RNase III-dependent mechanism. Genes Dev 1:1005-1013

Kuriki Y (1990) A nucleotide sequence in the translation start signal region is involved in heat shock-induced translation arrest in *E. coli*. FEBS Lett 264:121-124

La Teana A, Pon CL and Gualerzi CO (1993) Translation of mRNAs with degenerate initiation triplet AUU displays high IF2 dependence and is subject to IF3 repression. Proc Natl Acad Sci USA in press

Liao SM, Wu T, Chiang CH, Susskind MM and McClure WR (1987) Control of gene expression in bacteriophage P22 by a small antisense RNA. I. Characterization in vitro of the $P_{sar}$ promoter and the sar RNA transcript. Genes Dev 1:197-203

Lindahl L and Zengel JM (1986) Ribosomal genes in *E. coli*. Annu Rev Genet 20:297-326

Mahajna J, Oppenheim AB, Rattray A and Gottesman M (1986) Translation initiation of bacteriophage λ cII requires integration host factor. J Bacteriol 165:167-174

McCarthy JEG and Gualerzi CO (1990) Translational control of prokaryotic gene expression. Trends Genet 6:78-85

McPheeters DS, Stormo GD and Gold L (1988) Autogenous regulatory site on the bacterio-phage T4 gene 32 mRNA J Mol Biol 201:517-535

McMullin TW, Haffter P and Fox TD (1990) A novel small-subunit ribosomal protein of yeast mitochondria that interacts functionally with an mRNA-specific translation activator. Mol Cell Biol 10:4590-4595

Melançon P, Leclerc D and Brakier-Gingras L (1990a) A deletion mutant at the 5' end of *E. coli* 16S rRNA. Biochim Biophys Acta 1050:98-103

Melançon P, Leclerc D, Destroismaisons N and Brakier-Gingras L (1990b) The anti-Shine-Dalgarno region in *E. coli* 16S rRNA is not essential for the correct selection of transla-tional starts. Biochemistry 29:3402-3407

Mizuno T, Chou MY and Inouye M (1984) A unique mechanism regulating gene expression: Translational inhibition by a complementary RNA transcript (micRNA). Proc Natl Acad Sci USA 81:1966-1970

Nagai H, Yuzawa H and Yura T (1991) Interplay of two cis-acting mRNA regions in transla-tional control of $\sigma^{32}$ synthesis during the heat-shock response of *E. coli*. Proc Natl Acad Sci USA 88:10515-10519

Ohsawa H, Herrlich P and Gualerzi CO (1984) In vitro template activity of 0.3 mRNA from wild type and initiation mutants of bacteriophage T7. Mol Gen Genet 196:53-58

Petersen GB, Stockwell PA and Hill DF (1988) Messenger RNA recognition in *E. coli*: a possible second site of interaction with 16S rRNA. EMBO J 7:3957-3962

Philippe C, Portier C, Mougel M, Grunberg-Manago M, Ebel JP, Ehresmann B and Ehresmann C (1990) Target site of *E. coli* ribosomal protein S15 on its mRNA. Conformation and interaction with the protein. J Mol Biol 211:415-426

Philippe C, Eyermann F, Bernard L, Portier C, Ehresmann B and Ehresmann C (1993) Ribosomal protein S15 from *E. coli* modulates its own translation by trapping the ribosome on the mRNA initiation loading site. Proc Natl Acad Sci USA in press.

Ringquist S, Shinedling S, Barrick D, Green L, Binkley J, Stormo GD and Gold L (1992) Translation initiation in *E. coli*: sequences within the ribosome-binding site. Mol Microbiol 6:1219-1229

Romby P, Brunel C, Caillet J, Springer M, Grunberg-Manago M, Westhof E, Ehresmann C and Ehresmann B (1992) Molecular mimicry in translational control of *E. coli* threonyl-tRNA synthetase gene. Competitive inhibition in tRNA aminoacylation and operator repressor recognition switch using tRNA identity rules. Nucleic Acids Res 20:5633-5640

Rudd KE and Schneider TD, (1992) Compilation of *E. coli* ribosome binding sites. In: Miller JH (ed) A short course in bacterial genetics. Cold Spring Harbor Laboratory Press, Cold Spring Harbor, New York pp 17.19-17.45

Scherer GF, Walkinshaw MD, Arnott S and Morre DJ (1980) The ribosome binding sites recognized by *E. coli* ribosomes have regions with signal character in both the leader and protein coding segments. Nucleic Acids Res 8:3895-3907

Schneider TD and Stephens RM (1990) Sequence logos: a new way to display consensus sequences. Nucleic Acids Res 18:6097-6100

Schneider TD, Stormo GD, Gold L and Ehrenfeucht A (1986) Information content of binding sites on nucleotide sequences. J Mol Biol 188:415-431

Simons RW and Kleckner N (1983) Translational control of IS10 transposition. Cell 34:683-691

Spedding G, Gluick TC and Draper DE (1993) Ribosome initiation complex formation with the pseudoknotted α operon mRNA. J Mol Biol 229:609-622

Sprengart ML, Fatscher HP and Fuchs E (1990) The initiation of translation in *E. coli*: apparent base pairing between the 16S rRNA and downstream sequences of the mRNA. Nucleic Acids Res 18:1719-1723

Sussman JK and Simons RW (1993) *Escherichia coli* translation initiation factor 3 discriminates the initiation codon in vivo. Submitted for publication.

Takayama KM, Houba-Herin N and Inouye M (1987) Overproduction of an antisense RNA containing the oop RNA sequence of bacteriophage induces clear plaque formation. Mol Gen Genet 210:184-186

Tang CK and Draper DE (1989) Unusual mRNA pseudoknot structure is recognized by a protein translational repressor. Cell 57:531-536

Tang CK and Draper DE (1990) Evidence for allosteric coupling between the ribosome and repressor binding sites of a translationally regulated mRNA. Biochemistry 29:4434-4439

Vellanoweth RL and Rabinowitz JC (1992) The influence of ribosome binding-site elements on translational efficiency in *B. subtilis* and *E. coli* in vivo. Mol Microbiol 6:1105-1114

Winter RB, Morrissey L, Gauss P, Gold L, Hsu T and Karam J (1987) Bacteriophage T4 regA protein binds to mRNAs and prevents translation initiation. Proc Natl Acad Sci USA 84:7822-7826

Wulczyn FG, Boelker M and Kahmann R (1989) Translation of the bacteriophage Mu *mom* gene is positively regulated by the phage *com* gene product. Cell 57:1201-1210

Wu T, Liao SM, McClure WR and Susskind MM (1987) Control of gene expression in bacteriophage P22 by a small antisense RNA. II. Characterization of mutants defective in repression. Genes Dev 1:204-212

Wulczyn FG and Kahmann R (1991) Translational stimulation: RNA sequence and structure requirements for binding of com protein. Cell 65:259-269

# REGULATION OF PEPTIDE CHAIN TERMINATION

Yoshikazu Nakamura
Institute of Medical Science
University of Tokyo
P.O. Takanawa
Tokyo 108
Japan

Bacterial translation termination requires two codon-specific polypeptide release factors. The mechanism of stop codon recognition by release factors is unknown and holds considerable interest since it entails protein-RNA recognition rather than a codon-anticodon (RNA-RNA) interaction. Moreover, the stop codon is often recognized as alternate genetic codes for frameshifting, readthrough or selenocysteine incorporation (see review by Atkins et al., 1990). These programmed alternates in reading the stop codon require several regulatory elements to subvert the normal stop signal recognition in competition with the release factor. Therefore, translational termination can be now thought of in two ways: either as the general and fundamental event resulting in the release of the protein product, or as a pause or "yield" for more specialized events beyond the constraints of the genetic code (Nakamura et al., 1990; Tate & Brown, 1992). This short article summarizes our current understanding of the nature of the stop signal, the molecular basis of stop codon recognition, and its alternate readings, as well as the structural and functional organization of the release factor protein(s).

## I. The nature of the stop signal

*Hypothesis* : The termination signal in both prokaryotes and eukaryotes is a 4 base signal rather than a triplet, and the 4th base influences the effectiveness to specify a stop of protein synthesis.

Evolution of a protein-mediated response to a stop codon would not necessarily be restricted to a nonsense triplet; well characterized examples of other protein/nucleic acid recognition provide alternate models. Indeed, the increasing

NATO ASI Series, Vol. H 97
Post-transcriptional Control of Gene Expression
Edited by Orna Resnekov and Alexander von Gabain
© Springer-Verlag Berlin Heidelberg 1996

array of exceptions to the rule that a triplet codon signals "stop" with very high efficiency has led to reexamine the nature of the stop signal, and whether an efficient stop signal might be more than just the codons, UGA, UAA, and UAG. Tate and colleagues analyzed the region around the stop codons of over 800 genes from *E. coli* which revealed non randomness, particularly in the nucleotide following the triplet (Brown et al., 1990a). This non-randomness is dramatically accentuated in highly expressed genes which are terminated mainly by UAAU (56%) and UAAG (32%). This bias suggests that for efficient termination of protein synthesis, the stop signal is a tetranucleotide, and furthermore, the polypeptide release factors recognize this extended signal. The fourth base influences the efficiency of the signal. In support of this hypothesis, the rate of release factor-1 (RF-1) selection at four UAGN termination signals in bacteria varies 3 fold, with UAGU, being the best, and UAGA, the poorest with a correlation of 0.97 (Pederson & Curran, 1991). The translational pause would be shortest at those signals which select the RF more quickly.

Tate and colleagues have also analyzed the region around the stop signal in genes from a wide range of eukaryotes, including over 1300 human genes, and there is considerable suggestive evidence that the signal is also a tetranucleotide. For example, the preferred signal in highly expressed genes in Drosophila and *Saccharomyces cerevisiae* (organisms where rapid protein synthesis is an advantage) is UAAG (Brown et al., 1990b). The possibility of a four base eukaryotic stop signal had been foreshadowed by the observation that a minimum of a tetranucleotide was required for factor-mediated termination when measured in an *in vitro* assay, but this was interpreted as a requirement for codon binding to the ribosome rather than relating to its recognition by the release factor (Caskey, 1980).

## II. Release factor recognition of the stop signal

*Hypothesis* : The release factors recognize the stop signal as a double stranded structure, formed from an interaction of the mRNA with the rRNA at the 'decoding site'.

Translational stopping involves an intimate relationship between the ribosome (rRNA and proteins), the mRNA, and the release factor in response to the appearance of a stop, rather than a sense codon in the 'decoding' site of the ribosome. The fact that two of the release factors from bacteria exhibit codon specificity (RF-1: UAA, UAG; RF-2: UAA, UGA), suggests that they must interact directly with the codon. Until recently, evidence had been lacking for such a direct contact. Indeed, an alternative view has been proposed that there need not be direct contact between the factor and the stop codon, but rather that the codon may be decoded by rRNA at a site on the ribosome distant from the usual 'decoding site' for sense codons. The formation of a double stranded structure between the stop codon and the rRNA might cause a conformational change, allowing the release factors to bind to the ribosome. Protein-RNA crosslinking has provided evidence for a close contact between the stop codon and the release factor. A zero length crosslinker (a thiouracil residue in a stop codon within a designed RNA) was able to form a covalent complex with the release factor (Tate et al., 1990). This strategy identified Adenine 1408 of the 16S rRNA in the decoding site as the crosslinked rRNA residue (Tate et al., 1990). Residue 1408 forms part of a single stranded region in rRNA at the 'decoding site' in the sequence, Um$^5$CA, complementary to three positions with UGA, and in two positions with UAA and UAG. Interaction of the m$^5$C with the middle A of UAA and UAG is possible, however, since m$^5$C can base pair with A in the wobble position in at least one instance. Accordingly, one plausible model includes the recognition of a double stranded complex of codon and rRNA by the release factor at the decoding site. An alternative site for rRNA recognition of one of the stop codons, UGA, was suggested from the studies of Murgola and colleagues. It involves a region quite distant from the decoding site on the current models of the ribosome (Murgola et al., 1988). However, recent studies have suggested this region might in fact be close to the decoding site, and that it probably influences the reading of all three stop codons by affecting suppressor tRNA binding, not termination directly (Prescott et al., 1991). These results serve to illustrate that the mechanism of stop signal recognition by the release factors, whether direct or indirect, is still poorly defined.

## III. Genetic and functional organization of release factors

The genes encoding the *E. coli* release factors have been isolated. RF-1 was identified by a genetic screen for anti-suppression against an amber suppressor tRNA (Weiss et al., 1984), and RF-2, by an antibody-probed screen for protein overexpression in the Clarke-Carbon *E. coli* library (Caskey et al., 1984). The map position of RF-1 (designated *prfA*) is 27 min on the *E. coli* chromosome (Ryden et al., 1986) and RF-2 (designated *prfB*) is at 62 min in the same operon as the lysyl-tRNA synthetase gene (Kawakami et al., 1988b). Several mutants of RF-1 and RF-2 have been isolated and they often caused misreading of stop codons or frameshifting, as well as temperature-sensitive growth of the cells (Kawakami et al., 1988a; Elliott & Wang, 1991). Hence the reduced activity of release factors result in several translational errors *in vivo*, and these errors are likely caused by an abnormal pause of ribosomes at stop signals (Roesser et al., 1989). In *Salmonella typhimurium*, the *supK* gene encodes the *Salmonella* RF-2 protein, and one mutation, *supK584*, carries an opal UGA mutation within the coding region. We reasoned that the reduced intracellular level of RF-2 causes 'autogenous suppression' of the opal RF-2 mutation (Kawakami & Nakamura, 1990). Therefore, it is now evident that changing either the concentration or the activity of the release factor generates altered reading of the stop signal.

In *E. coli*, a third factor, RF-3, is known to stimulate the activities of RF-1 and RF-2 and binds guanine nucleotides, but is not codon-specific (Goldstein & Caskey, 1970). However, this factor has received little attention during two decades until recently. We selected pseudo-revertants from the temperature-sensitive RF-2 mutant to search for cellular components, including RF-3, that are involved in release factor activity (Mikuni & Nakamura, manuscript in preparation). One class of such mutations, named *srbA*, was mapped to the 99 min region on the *E. coli* chromosome. In parallel to this suppressor-gene hunting experiment, other trials were made to select nonsense suppressor mutations by transposon-mediated mutagenesis by assuming that, if transposon insertion nullifies a gene encoding a factor that stimulates stop-codon recognition, those mutants may exert nonsense suppression. In fact, one such mutation was isolated and named *tos* for transposon mediated opal suppressor (Mikuni & Nakamura, manuscript in preparation). Surprisingly, the *tos* mutation affected the same gene as *srbA* and caused suppression of all three stop codons. Moreover, the *tos* protein has guanine nucleotide-binding motifs and shares significant sequence homology with elongation factor EF-G. Taken these and other results into consideration, it is very likely that we have succeeded to identify the RF-3 gene beyond two decades of silence.

The primary sequences for several bacterial release factors and the yeast mitochondrial factor have been determined (Craigen et al., 1985; Kawakami et al., 1988b; Kawakami & Nakamura, 1990; Pel et al., 1992). Revised sequence comparison between RF-1 and RF-2 predicts three well conserved domains (domains I through III) (Mikuni et al., 1991). Three amino acid substitutions in RF-2, both in conserved (domain II) and nonconserved regions (between domains I and II, and the C-terminal), lost the normal level of UGA recognition (Mikuni et al., 1991).

The C-terminal mutation conferred temperature-sensitive lethality to the cell. In *Salmonella typhimurium*, it is also known that mutations in domain II generate UGA suppression (Elliott & Wang, 1991). As yet, there is no clear definition of which of the several functional activities reside in which part of the molecules. However, preliminary studies have been also undertaken to identify functionally and structurally important domains of the two bacterial release factors, by creating chimeras between the two genes, and have succeeded in producing factors which have lost or retained codon recognition activity independent of hydrolytic function (Moffat et al., 1993; Nobukuni & Nakamura, unpublished). One expressed chimeric protein has retained recognition activity but is defective in peptidyl-tRNA hydrolysis, while another has lost codon recognition but retained some peptidyl-tRNA hydrolysis activity (Moffat et al., 1993). From these studies, it has been proposed that sequences in the N terminal region and the highly conserved region towards the C terminus of the protein contribute to the 'hydrolysis' domain, whereas sequences from the middle of the protein and an extreme C terminal region contribute to the codon recognition domain. One hybrid protein has only 10 RF-2 residues in an RF-1 background and yet is defective in peptidyl-tRNA hydrolysis (Moffat et al., 1993). A motif involving basic residues has been identified in a kinase that interacts with double stranded RNA in an interferon system as a putative RNA binding domain. The bacterial release factors have similar motifs just at the edge of the highly conserved region within that segment of the primary sequence which likely contribute to codon recognition. Besides the genetic approach, the successful crosslinking of a thiU residue within a stop codon to the release factor (Tate et al., 1990) also provides a method for identifying amino acid residues within the codon recognition domain of the factor.

A putative mammalian release factor sequence has been obtained, but it seems to be identical to tryptophanyl-tRNA synthetase, and to have no homology with other known factors (Lee et al., 1990). This raises the intriguing possibility that the mammalian factor is a single protein with two functions in protein synthesis, and that it has evolved independently from those factors mediating the stop of protein synthesis in bacteria. However, Tate and others (personal communication) have obtained conflicting evidence regarding the protein's activities, and it is likely that the protein has been misidentified as a release factor (or they are closely related proteins being not identical).

## IV. Alternate readings of the stop signal

The failure of certain stop codons to signal a complete stop in protein synthesis indicates that in certain contexts they may act as weak stop signals, at least when there are possible alternative events. In most cases, the events that compete for the stop of protein synthesis are enhanced by other factors such as the secondary or tertiary structure of the mRNA. Presumably, there is a complex kinetic balance between termination and the competing process, whether it is suppression, frameshifting or specific amino acid incorporation at these sites.

Perhaps the most dramatic example of such an event is the use of the UGA codon to encode for selenocysteine (Stadman, 1991). This "misreading" of the UGA codon occurs in mRNAs for the human proteins, glutathione peroxidase, 5'monodeiodinase, which process the thyroid hormone, and SelP, which contains 10 UGAs in frame, as well as the *E. coli* formate dehydrogenase (Berry et al., 1990a; 1990b; Zinoni et al., 1987). Why do not these UGAs signal stop efficiently, like the majority of their compatriots? The UGAs that code for selenocysteine in mammalian genes are usually found in the sequence UGAC/A. This sequence is predicted to be within the group of poor stop signals since they are underutilized in highly expressed genes (Brown et al., 1990b). In *E. coli*, it is of interest that selenocysteine is incorporated into formate dehydrogenase at a UGA(C) signal, and that this 4 base sequence is found very rarely at natural termination sites.

It is now clear that a downstream stem loop structure is a crucial element for the incorporation of selenocysteine into formate dehydrogenase in *E. coli* (Zinoni et al., 1990), and into mammalian glutathione peroxidase, 5'monodeiodinase and SelP (Berry et al., 1993). Böck and colleagues have beautifully demonstrated that a specific tRNA, with a UCA anticodon, encoded by the *selC* gene is charged with selenocysteine and is used for alternate reading of UGA in association with the *selB* protein in *E. coli* (Leinfelder et al., 1988; Forchhammer et al., 1989). SelB is a novel translation factor homologous to elongation factor EF-Tu and binds specifically both the selenocysteine-charged *selC*-tRNA and the stem-loop structure of RNA immediately downstream of UGA, thereby specifying the decoding site. On the other hand, this regulatory hairpin structure of RNA is located far downstream of UGAs in mammalian selenoprotein genes and the mechanism of selenocysteine decoding in eukaryotes is not known.

The UGA(C) sequence, a rare signal for natural stop, is also found at the frameshift site in the bacterial gene for RF-2, another bacterial site where there is a well characterized failure to terminate efficiently (Craigen et al., 1985). The RF-2 coding region contains a premature (in frame) UGA stop at codon position 26 and +1 frameshift occurs before this stop signal to complete synthesis of RF-2. Weiss et al. (1988) have demonstrated that an internal Shine-Dalgarno sequence immediately upstream of UGA is necessary for frameshifting by base-pair formation with the 3' end of the 16S rRNA that probably leads to ribosome stalling during elongation. This is a natural mechanism of autotranslational feedback control. In fact, the termination signal can be strengthened *in vitro* by increasing the concentration of RF-2 or weakened *in vivo* by reducing the activity of RF-2, despite there being other context elements surrounding the site which favor frameshifting (Donly et al., 1990; Mikuni et al., 1991). Perhaps the most bizarre event yet reported is the bypassing of 50 nucleotides by a ribosome in the translation of T4 DNA topoisomerase mRNA. Here, an in phase UAGC, also rare at natural termination sites, is part of an elaborate mechanism to achieve this bypass (Weiss et al., 1990).

Termination of protein synthesis has been traditionally referred to as the hydrolysis of peptidyl-tRNA causing the release of the completed protein from the ribosome. However, as described above, a translational stop can be thought of in two ways: either as a complete stop, mediated by the peptide-chain release factor, or a pause for more specialized events beyond the normal constraints of the genetic code. This pause may allow for a diverse array of possible competing events to occur at the stop signal, i.e., alternate readings of the genetic code. This raises two important questions: how are stop codons recognized as a signal for the termination of protein synthesis, and how is this recognition subverted to specify the 'alternative genetic codes'? Obviously, further studies are necessary to uncover the molecular basis for stop-codon recognition, still a long standing and perhaps the last remaining coding problem, and the alternate genetic code, the new frontier to understand coding, in translation.

I thank Dr. Warren P. Tate for cooperation in the Human Frontier Science Program application (awarded in 1993; principle investigator Y.N.) that was in part adopted to this article.

## References

Atkins JF, Weiss RB, Gesteland RF (1990) Ribosome gymnastics - degree of difficulty 9.5, style 10.0. Cell 62:413-423
Berry MJ, Banu L, Larsen PR (1991a) Type I idiothyronine deiodinase is a selenocysteine-containing enzyme. Nature 349:438-440

Berry MJ, Banu L, Chen Y, Mandel S, Kieffer JD, Harney JW, Larsen PR (1991b) Recognition of UGA as a selenocysteine codon in Type I deiodinase requires sequences in the 3' untranslated region. Nature 353:273-276

Berry MJ, Banu L, Harney JW, Larsen PR (1993) Functional characterization of the eukaryotic SECIS elements which direct selenocysteine insertion at UGA codons. EMBO J 12:3315-3322

Brown CM, Stockwell PA, Trotman CNA, Tate WP (1990a) The signal for the termination of protein synthesis in prokaryotes. Nucleic Acids Res 18:2079-2086

Brown CM, Stockwell PA, Trotman CNA, Tate WP (1990b) Sequence analysis suggests that tetranucleotides signal the termination of protein synthesis in eukaryotes. Nucleic Acids Res 18:6339-6345

Caskey CT (1980) Peptide chain termination. Trends Biochem Sci 5:234-237

Caskey CT, Forrester WC, Tate W, Ward CD (1984) Cloning of the *Escherichia coli* release factor 2 gene. J Bacteriol 158:365-368

Craigen WJ, Cook RG, Tate WP, Caskey CT (1985) Bacterial peptide chain release factors: conserved primary structure and possible frameshift regulation of release factor 2. Proc Natl Acad Sci USA 82:3616-3620

Donly BC, Edgar CD, Adamski FM, Tate WP (1990) Frameshifting at the internal stop codon within the mRNA for bacterial release factor 2: partly functional mutants result in frameshift enhancement. Nucleic Acids Res 18:6517-6522

Elliott T, Wang X (1991) *Salmonella typhimurium prfA* mutants defective in release factor 1. J Bacteriol 173:4144-4154

Forchhammer K, Leinfelder W, Böck A (1989) Identification of a novel translation factor necessary for the incorporation of selenocysteine into protein. Nature 342:453-456

Goldstein JL, Caskey CT (1970) Peptide chain termination: effect of protein S on ribosomal binding of release factors. Proc Natl Acad Sci USA 67:537-543

Kawakami K, Inada T, Nakamura Y (1988a) Conditionally lethal and recessive UGA-suppressor mutations in the *prfB* gene encoding peptide chain release factor 2 of *Escherichia coli*. J Bacteriol 170:5378-5381

Kawakami K, Jönsson YH, Björk GR, Ikeda H, Nakamura Y (1988b) Chromosomal location and structure of the operon encoding peptide-chain-release factor 2 of *Escherichia coli*. Proc Natl Acad Sci USA 85:5620-5624

Kawakami K, Nakamura Y (1990) Autogenous suppression of an opal mutation in the gene encoding peptide chain release factor 2. Proc Natl Acad Sci USA 87:8432-8436

Lee CC, Craigen WJ, Muzny DM, Harlow E, Caskey CT (1990) Cloning and expression of a mammalian chain release factor with sequence similarity to tryptophanyl-transfer RNA synthetases. Proc Natl Acad Sci USA 87:3508-3512

Leinfelder W, Zehelein E, Mandrand-Berthelot MA, Böck A (1988) Gene for a novel tRNA species that accepts L-serine and cotranslationally inserts selenocysteine. Nature 331:723-725

Mikuni O, Kawakami K, Nakamura Y (1991) Sequence and functional analysis of mutations in the gene encoding peptide-chain-release factor 2 of *Escherichia coli*. Biochimie 73:1509-1516

Moffat JG, Donly BC, McCaughan KK, Tate WP (1993) Functional domains in the Escherichia coli release factors activities of hybrids between RF-1 and RF-2. Eur J Biochem 231:749-756

Murgola EJ, Hijiazi KA, Goringer HU, Dahlberg AE (1988) Mutant 16S rRNA: a codon specific translational suppressor. Proc Natl Acad Sci USA 85:4162-4165

Nakamura Y, Kawakami K, Mikuni O (1990) Alternative translation and functional diversity of release factor 2 and lysyl-tRNA synthetase. In: Post-Transcriptional Control of Gene Expression (McCarthy et al., eds), pp. 455-464, Elsevier Science Publishing Co, Inc, New York

Pederson WT, Curran JF (1991) Effects of the nucleotide 3' to an amber codon on ribosomal selection rates of suppressor tRNA and release factor-1. J Mol Biol 219:231-241

Pel HJ, Rep M, Grivell LA (1992) Sequence comparison of new prokaryotic and mitochondrial members of the polypeptide chain release factor family predicts a five-domain model for release factor structure. Nucleic Acids Res 20:4423-4428

Prescott CD, Krabben L, Nierhaus K (1991) Ribosomes containing the C1054 deletion mutation in *E. coli* act as suppressors at all three nonsense codons. Nucleic Acids Res 19:5281-5283

Roesser JR, Nakamura Y, Yanofsky C (1989) Regulation of basal level expression of the tryptophane operon of *E. coli*. J Biol Chem 264:12284-12288

Ryden M, Murphy J, Martin R, Isaksson L, Gallant J (1986) Mapping and complementation studies of the gene for release factor 1. J Bacteriol 168:1066-1069

Stadman TC (1991) Biosynthesis and function of selenocysteine-containing enzymes. J Biol Chem 266:16257-16260

Tate WP, Brown CM (1992) Translational termination: "stop" for protein synthesis or "pause" for regulation of gene expression. Biochemistry 31:2443-2450

Tate W, Greuer B, Brimacombe R (1990) Codon recognition in polypeptide chain termination: site directed crosslinking of termination codon of *Escherichia coli* release factor 2. Nucleic Acids Res 18:6537-6544

Weiss RB, Dunn DM, Dahlberg AE, Atkins JF, Gesteland RF (1988) Reading frame switch caused by base-pair formation between the 3' end of 16S rRNA and the mRNA during elongation of protein synthesis in *Escherichia coli*. EMBO J 7:1503-1507

Weiss RB, Huang WM, Dunn DM (1990) A nascent peptide is required for ribosomal bypass of the coding gap in bacteriophage T4 gene 60. Cell 62:117-126

Weiss RB, Murphy JP, Gallant JA (1984) Genetic screen for cloned release factor genes. J Bacteriol 158:362-364

Zinoni F, Birkmann A, Leinfelder W, Böck A (1987) Cotranslational insertion of selenocysteine into formate dehydrogenase from *Escherichia coli* directed by a UGA codon. Proc Natl Acad Sci USA 84:3156-3160

Zinoni F, Heider J, Böck A (1990) Features of the formate dehydrogenase mRNA necessary for decoding of the UGA codon as selenocysteine. Proc Natl Acad Sci USA 87:4660-4664

# STRUCTURE-FUNCTION CORRELATIONS IN THE BACTERIAL RIBOSOME

Richard Brimacombe
Max-Planck-Institut für Molekulare Genetik
Ihnestrasse 73
1000 Berlin 33, Germany

The bacterial ribosome consists of three RNA molecules totalling over 4500 nucleotides, and about 55 different protein molecules containing a total of over 7300 amino acids. Any detailed unterstanding at the molecular level of the function of this complex particle will obviously require a correspondingly detailed knowledge of its structure, and the purpose of this article is to briefly summarize the current status of these structure-function correlations in the case of the best-studied ribosome, namely that from *Escherichia coli*.

The ribosome is of a very awkward size for structural studies. On the one hand, it is too small to be examined with sufficient accuracy by electron microscopy, but on the other hand it is still too large and its structure too irregular for an analysis at atomic resolution by X-ray crystallography (although progress is being made in this direction (Von Böhlen et al, 1991)). The three-dimensional structure therefore has to be put together piece by piece, using information from a variety of techniques and sources, with varied levels of resolution. Since the protein moiety consists of many separate molecules (named S1 to S21 in the small 30S ribosomal subunit of *E. coli*, and L1 to L36 in the large 50S subunit), it was possible from an early stage to carry out low-resolution studies on the topographical arrangement of the proteins, without necessarily needing to know the individual amino acid sequences. In contrast, the mere fact that the RNA moiety consists primarily of a single large molecule in each subunit (1542 nucleotides long in the 16S RNA

NATO ASI Series, Vol. H 97
Post-transcriptional Control of Gene Expression
Edited by Orna Resnekov and Alexander von Gabain
© Springer-Verlag Berlin Heidelberg 1996

of the small subunit, and 2904 nucleotides long in the 23S RNA of the large subunit, the latter also containing the 120 nucleotide-long 5S RNA) has the consequence that very little can be done in the way of structural or topographical investigation without reference to its primary sequence. However, for the same reason, once the RNA nucleotide sequences became available, the structural studies on the RNA tended a priori to be made at a higher level of resolution (i.e. at the nucleotide level) than those with the ribosomal proteins. Thus, whereas much of the low-resolution data on the proteins has been available for quite some time, the upsurge in high-resolution data on the RNA has been rapid and recent, and also includes a considerable amount of information relating to the ribosomal function.


ARRANGEMENT OF RIBOSOMAL PROTEINS

The three-dimensional distribution of the *E. coli* ribosomal proteins has been studied by several different methods, and the results have been the subject of a number of reviews (e.g. Wittmann, 1983). The three most informative techniques have been protein-protein cross-linking, immuno electron microscopy, and low-angle neutron scattering. In the first of these, protein-protein cross-links are induced *in situ* in the ribosomal subunits by treatment with an appropriate bifunctional reagent, the cross-linked protein pairs are isolated, and their components identified either by gel electrophoresis (after reversing the cross-link) or by immunological analysis. The immuno electron microscopic method makes use of the fact that the ribosomal subunits have very characteristic shapes in the electron microscope; the 30S subunit consists of a "head", a "body" and a large "lateral protuberance" which is separated from the body by a "cleft", whereas the 50S subunit consists of a "body" surmounted by three distinct protuberances. Antibodies specific to the individual ribosomal proteins can be bound to the corresponding subunit, and their positions on these characteristic shapes assessed by electron microscopy.

In the third method - low-angle neutron scattering - ribosomal subunits containing two deuterated proteins in a matrix of "normal" hydrogen-containing proteins (or vice versa) can be prepared by reconstitution of 30S or 50S subunits from their constituent RNA and protein components, and the neutron scattering properties of these reconstituted particles are used to compute the distance between the centres of mass of the two proteins chosen. It is important to note that, while the latter method measures distances between protein mass centres, the cross-linking and electron microscopy methods provide data relating to points on the surface of the protein concerned. Since the proteins are of unknown shapes and dimensions this introduces a corresponding element of uncertainty when comparing the results from the different techniques, but it can be said that - within the limits of this uncertainty - there is a generally acceptable level of agreement between the various data sets. Particularly important has been the establishment by the neutron scattering method of a three-dimensional map of the mass centres of all 21 proteins of the 30S subunit (Capel et al, 1988).

FOLDING THE RIBOSOMAL RNA

The first step towards determining the three-dimensional structure of the large 16S and 23S RNA molecules was to establish their secondary structures, and this was achieved primarily by the application of the sequence comparison approach. The primary sequence of ribosomal RNA has been highly conserved throughout evolution, and therefore sequence variations between one organism and another can be used to test putative secondary structural elements; a base change in one strand of a double-helical element must be accompanied by a "compensating" change in the opposing strand (if the proposed structure is correct) in order to preserve the base-pairing (e.g. Brimacombe, 1984). Since a very large number of ribosomal RNA sequences is now available, the secondary structures derived in this way for both the 16S and 23S molecules are supported by literally thousands of such pairs of compensating base changes (e.g. Noller, 1984), and these structures provide a firm basis for the further folding of the RNA.

The sequence comparison approach can also be used to locate tertiary base-pairings within the ribosomal RNA, and several important interactions have been found in this way. However, more extensive data sets relating to tertiary contacts in the RNA have been obtained by intra-RNA cross-linking experiments (e.g. Döring et al, 1991). As with the protein-protein cross-linking method described above, these cross-links are induced *in situ* within the intact ribosomal subunits, and then after deproteinization the sites involved on the RNA are analysed as precisely as possible. Combined with the localization on the ribosomal subunit surfaces by immuno electron microscopy of several specific sites in the RNA molecules (5' or 3' termini, or modified nucleotides (Gornicki et al, 1984)), these data begin to constrain the RNA into the compact three-dimensional shape which it occupies within the ribosome.

FITTING THE PROTEINS TO THE RNA

"Connecting links" between the RNA and the ribosomal proteins have been obtained by two methods. The first of these is again a cross-linking method - in this case *in situ* RNA-protein cross-linking - which has been used to establish a large number of contacts between individual proteins and specific sites on both the 16S and 23S RNA (Brimacombe, 1991). The second method is protein foot-printing, whereby the accessibility of the RNA to chemical or nuclease attack is examined in the presence or absence of the protein under study; this technique has also provided an extensive data set, mostly relating to the proteins from the 30S subunit (Stern et al, 1988).

The RNA-protein cross-linking or foot-printing data can be combined with the established arrangement of the ribosomal proteins, so as to constrain the secondary structure of the RNA into a relatively compact and detailed three-dimensional structure. Such model structures have been proposed both for the 16S RNA in the 30S subunit (Brimacombe et al, 1988; Stern et al, 1988) and for the 23S RNA in the 50S subunit (Mitchell et al, 1990). These models appeared to be able to accommodate the great majority of the available experimental data, and showed for example that the most highly conserved regions of the 16S RNA secondary structure were concentrated into a belt around the middle of the 30S subunit, whereas the corresponding regions of the 23S RNA were concentrated on the 30S-50S interface side of the 50S subunit.

FUNCTIONAL SITES ON THE RNA

The foot-printing technique mentioned above has been extensively applied to locate sites on the ribosomal RNA that are involved in the binding of functional ligands such as tRNA (Moazed and Noller, 1990) or antibiotics which interfere with various steps in the protein biosynthetic process (Moazed and Noller, 1987). In addition, a number of specific mutations in the RNA have been identified which cause resistance to certain of these antibiotics (e.g. Powers and Noller, 1990; Vester and Garrett, 1988). These sites are all spread widely in the primary sequence or secondary structure of the 16S and 23S RNA, but - in contrast - in the three-dimensional models they are grouped into distinct clusters. In the model of the 50S subunit (Mitchell et al, 1990) this functional cluster is at the base of the central protuberance, where the peptidyl transferase centre has been by located electron microscopic studies (Wittmann, 1983). On the other hand, in the 30S models, the functional sites are grouped into two clusters, one at the base of the cleft, where electron microscopy has identified the decoding site, and the other on the opposite side of the subunit. It was reasoned (Stiege et al, 1988; Stern et al, 1988) that the latter group of sites must be involved in the function in an allosteric manner, by virtue of the relatively large distance separating them from the "cleft group".

The inability of the foot-printing technique to distinguish between direct and allosteric effects is a serious disadvantage of the method, and a second disadvantage is that foot-printing can give no information relating to the orientation of the ligand (such as tRNA) being studied. However, both of these drawbacks can be overcome by application of the "site-directed cross-linking" method, in which photo-labels are introduced at any desired position in the ligand of interest, and, after binding to the ribosome and photo-activation, the sites of attachment to the ribosomal RNA (or proteins) are analysed. The site-directed cross-linking method is being used in our laboratory to study the orientations on the ribosome of mRNA, tRNA and the nascent peptide chain. Most importantly, a series of experiments with mRNA analogues has given clear indication (Dontsova et al, 1992) that the models for the 30S subunit described above are incorrect in some important aspects; photo-labels at positions only four or five bases apart on the mRNA became attached to sites on the 16S RNA that are widely separated in the models, and these mRNA cross-links were observed under a range of different conditions. Furthermore, since publication of the 30S models, a number of other lines of evidence have begun to suggest that the separation of the functional sites into two groups within the 30S subunit as discussed above, was not a valid conclusion (Brimacombe, 1992). It seems likely that the low resolution inherent in the placement of RNA elements in relation to the positions of the ribosomal proteins (cf. the discussion above in the section on "Arrangement of Ribosomal Proteins") has led to cumulative errors in parts of the 16S models, and that this factor - combined with possible misinterpretations of some of the older electron microscopy data (Brimacombe, 1992) - is responsible for the discrepancy in the locations of the functional sites. At all events, some revision of the RNA models is clearly necessary.

THE NEW MODEL-BUILDING STRATEGY

Recently, a new electron microscopic model of the 70S ribosome has been published (Frank et al, 1991), based on three-dimensional image reconstruction of micrographs of ribosomes in amorphous ice (i.e. without negative staining). This recon-struction shows a number of new features - such as a "bridge" between the 30S and 50S subunits - which fit well to the biochemical observations. Furthermore, it also shows a cavity between the two subunits which can rather precisely accommodate two molecules of tRNA. Accordingly, we propose to build up a model of those elements of the 16S or 23S ribosomal RNA that are in contact with the tRNA-mRNA-peptide complex, by using the site-directed cross-linking data (obtained with tRNA, mRNA or the nascent peptide) as the "primary model-building criterion". Other regions of the ribosomal RNA will be automatically brought into this model, either by virtue of their neighboured location within the secondary structure, or on the basis of inter- or intra-RNA cross-linking data. In this way we expect to be able to model the regions of ribosomal RNA forming the periphery of the interface cavity in the model of Frank et al (1991), and then to extend the structure to other parts of the ribosome. The RNA-protein cross-linking and foot-printing data will be used as a "secondary criterion", that is to check the general locations of the corresponding RNA regions relative to the protein arrangement, especially in those regions more remote from the active centre (the tRNA-mRNA complex).

Needless to say, the new model is still far from complete. However, the data so far available show a number of encouraging features. For example, 27 of the 29 modified nucleotides in the *E. coli* ribosomal RNA have been shown to form a single cluster surrounding the mRNA-tRNA complex (Brimacombe et al, 1993), although these nucleotides are widely scattered throughout the 16S and 23S molecules. Similarly, a number of widely-separated sites in the 16S RNA implicated in the binding of the antibiotics spectinomycin and streptomycin are now also clustered around the anticodon loop of the tRNA at the ribosomal A-site. Last but not least, the geometry of the RNA elements involved seems to fit very well to the dimensions of the interface cavity.

REFERENCES

Brimacombe R (1984) Conservation of structure in ribosomal RNA. Trends Biochem Sciences 9:273-277
Brimacombe R, Atmadja J, Stiege W, Schüler D (1988) A detailed model of the three-dimensional structure of *E. coli* 16S ribosomal RNA *in situ* in the 30S subunit. J Mol Biol 199:115-136
Brimacombe R (1991) RNA-protein interactions in the *E. coli* ribosome. Biochimie 73:927-936
Brimacombe R (1992) Structure-function correlations (and discrepancies) in the 16S ribosomal RNA from *E. coli*. Biochimie 74:319-326
Brimacombe R, Mitchell P, Osswald M, Stade K, Bochkariov D (1993) Clustering of modified nucleotides at the functional center of bacterial ribosomal RNA. FASEB J, in press
Capel MS, Kjeldgaard M, Engelman DM, Moore PB (1988) Positions of S2, S13, S16, S17, S19 and S21 in the 30S ribosomal subunit of *E. coli*. J Mol Biol 200:65-87
Döring T, Greuer B, Brimacombe R (1991) The three-dimensional folding of ribosomal RNA; localization of a series of intra-RNA cross-links in 23S RNA induced by treatment of *E. coli* 50S ribosomal subunits with bis-(2-chloroethyl)-methylamine. Nucleic Acids Res 19:3517-3524
Dontsova O, Dokudovskaya S, Kopylov A, Bogdanov A, Rinke-Appel J, Jünke N, Brimacombe R (1992) Three widely separated positions in the 16S RNA lie in or close to the ribosomal decoding region; a site-directed cross-linking study with mRNA analogues. EMBO J 11:3105-3116
Frank J, Penczek P, Grassucci R, Srivastava S (1991) Three-dimensional reconstruction of the 70S E. coli ribosome in ice; the distribution of ribosomal RNA. J Cell Biol 115:597-605
Gornicki P, Nurse K, Hellman W, Boublik M., Ofengand J (1984) High resolution localization of the tRNA anticodon inter-action site on the *E. coli* 30S ribosomal subunit. J Biol Chem 259:10493-10498
Mitchell P, Osswald M, Schüler D, Brimacombe R (1990) Selective isolation and detailed analysis of intra-RNA cross-links induced in the large ribosomal subunit of *E. coli*; a model for the tertiary structure of the tRNA binding domain in 23S RNA. Nucleic Acids Res 18:4325-4333
Moazed D, Noller HF (1987) Interaction of antibiotics with functional sites in 16S ribosomal RNA. Nature 327:389-394
Moazed D, Noller HF (1990) Binding of tRNA to the ribosomal A and P sites protects two distinct sets of nucleotides in 16S rRNA. J Mol Biol 211:135-145
Noller HF (1984) Structure of ribosomal RNA. Annu Rev Biochem 53:119-162

Powers T, Noller HF (1991) A functional pseudoknot in 16S ribosomal RNA. EMBO J 10:2203-2214

Stern S, Weiser B, Noller HF (1988) Model for the three-dimensional folding of 16S ribosomal RNA. J Mol Biol 204:447-481

Stiege W, Stade K, Schüler D, Brimacombe R (1988) Covalent cross-linking of poly(A) to *E. coli* ribosomes, and localization of the cross-link site within the 16S RNA. Nucleic Acids Res 16:2369-2388

Vester B, Garrett RA (1988) The importance of highly conserved nucleotides in the binding region of chloramphenicol at the peptidyl transfer centre of *E. coli* 23S ribosomal RNA. EMBO J 7:3577-3587

Von Böhlen K, Makowski I, Hansen HAS, Bartels H, Berkovitch-Yellin Z, Zaytzev-Bashan A, Meyer S, Paulke C, Franceschi F, Yonath A (1991) Characterization and preliminary attempts for derivatization of crystals of large ribosomal subunits from *H. marismortui* diffracting to 3 Å resolution. J Mol Biol 222:11-15

Wittmann HG (1983) Architecture of prokaryotic ribosomes. Annu Rev Biochem 52:35-65

# INITIATION FACTOR-DEPENDENT EXTRACTS: A TOOL TO STUDY TRANSLATION INITIATION IN EUKARYOTES

Hans Trachsel, Michael Altmann and Sylviane Blum
Institute of Biochemistry
and Molecular Biology
University of Berne,
Bühlstrasse 28
3012 Berne, Switzerland

## Introduction

Several approaches are available to create initiation factor-dependent extracts from *Saccharomyces cerevisiae* cells. Temperature-sensitive mutations in initiation factor genes can be generated and lysates from such strains prepared. Heating of the extract to the non-permissive temperature leads to inactivation of the factor *in vitro* and results in a cell-free protein synthesizing system whose activity is dependent on exogenous initiation factor. This method was used to prepare eIF-4E- and Prt1-dependent extracts. Alternatively, the gene encoding an initiation factor can be cloned behind the GAL1/10 promoter and cells containing this construct transferred to galactose-containing medium. This leads to shut-off of initiation factor synthesis *in vivo*. Lysates prepared from cells which slow down growth in galactose-containing medium are dependent on addition of the initiation factor for translation *in vitro*. This approach was used to create an eIF-4A-dependent extract. Initiation factor-dependent extracts allow the study of a single component in an otherwise complete cell-free system. They are useful to determine the factor requirements for translation of different mRNAs and to study structure-function relationships in individual initiation factors.

NATO ASI Series, Vol. H 97
Post-transcriptional Control of Gene Expression
Edited by Orna Resnekov and Alexander von Gabain
© Springer-Verlag Berlin Heidelberg 1996

## Prt1-dependent extract

The gene encoding Prt1 was cloned by complementation of the temperature-sensitive (ts) phenotype of a *S. cerevisiae* mutant strain (Keierleber et al., 1986; Hanic-Joyce et al., 1987). It was shown to most likely encode a translation initiation factor. The protein was never purified but the sequence analysis predicts that Prt1 is a 80 kD protein. It is essential for viability and stimulates Met-tRNA binding to 40S ribosomal subunits. Extracts prepared from prt1 mutants are active in *in vitro* translation at 23°C but inactive after a preincubation for 5-10 min at 37°C. Partially purified preparations of Prt1 can restore translational activity of these extracts. This proves that Prt1 is a translation factor and creates an assay to purify the protein from *S.cerevisiae* cells. Furthermore, this system should be very valuable to obtain more detailed information about the function of Prt1 in translation initiation.

## eIF-4E-dependent extract

Yeast eIF-4E was isolated by affinity chromatography on m7GTP-agarose columns. The cDNA encoding eIF-E was isolated from a lambda gt11 library using a polyclonal anti-eIF-4E antibody (Altmann et al., 1987). To study the requirement of different mRNAs for eIF-4E we prepared an eIF-4E-dependent translation system. To create a ts mutation in the eIF-4E gene we randomly mutagenized it and introduced into a yeast cell which had its chromosomal copy of the eIF-4E gene deleted and carried a wild-type copy of the gene under the control of the GAL1/10 promoter on a plasmid. Ts mutants were identified by growing transformants at 23°C and replica plating and incubation at 37°C. One ts mutant strain, ts 3-2, was further analyzed. The cells of this strain grow well at 23°C but stop growing at 37°C. Cell-free system prepared from this strain are dependent on exogenous eIF-4E for translation of total yeast mRNA after heating them for 5-10 min. at 37°C (Altmann et al., 1989)(Fig. 1).

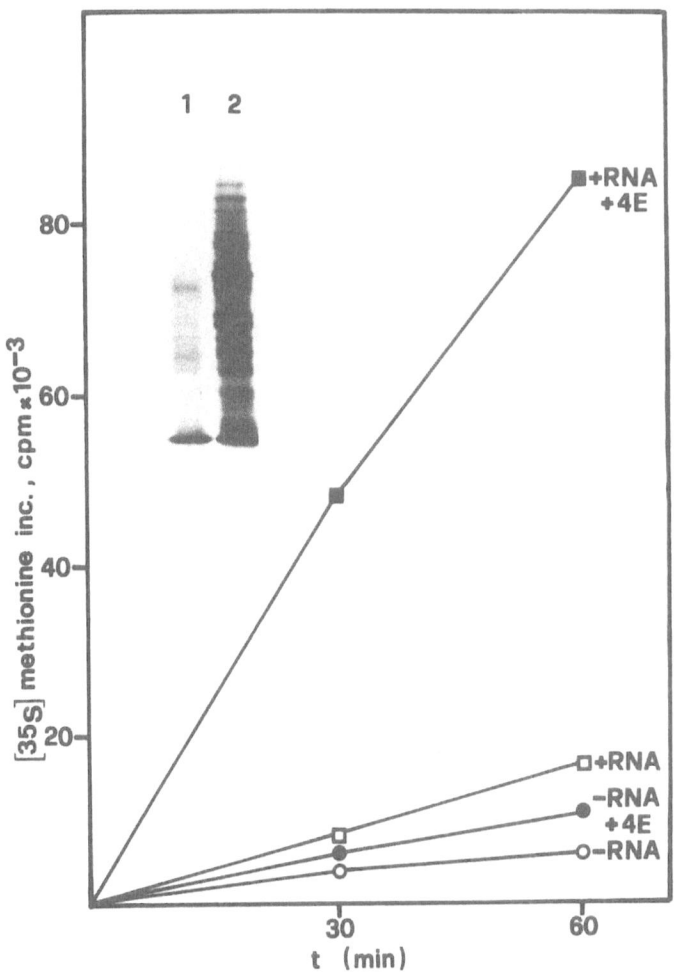

Fig. 1: Cell-free translation in eIF-4E-dependent extract. Initiation factor 4E-dependent extract was digested with micrococcal nuclease, reaction mixtures (20 μl) were incubated at 37°C for 10 min in the absence of labeled amino acid and mRNA, and translation was started by addition of [35S]methionine and (where incubated) 10 μg of total yeast RNA. Inset: In vitro translation products after 60 min of incubation. Lane 1, minus eIF-4E; lane 2, plus 50 ng of eIF-4E.

Using this system to translate a number of different mRNAs it could be shown that mRNAs with little secondary structure in their 5' untranslated region require no exogenous factor for translation.

## eIF-4A-dependent extract

Initiation factor 4A (eIF-4A) is a 43 kilodalton protein. It is required for mRNA binding to ribosomes and ATP-hydrolysis-dependent RNA secondary structure unwinding during the scanning process. We isolated yeast eIF-4A, and constructed an eIF-4A-dependent cell-free translation system. This was achieved by cloning the gene behind the GAL1/10 promoter and intoducing this construct into a yeast strain having both eIF-4A genes interrupted. The resulting transformants grow on galactose-containing but not on glucose-containing media. Transfer of cells into glucose-containing medium results in slow down of growth after about 8h at 25°C. Cell-free systems prepared from cells at this stage are strongly dependent on exogenous eIF-4A for translation of total yeast mRNA (Blum et al., 1989)(Fig. 2). We used this system and other assays to elucidate the function of an alanine residue in the putative ATP binding site conserved in all RNA helicases of the DEAD box protein family found so far (Linder et al., 1989). Factor eIF-4A was mutated and expressed in *Escherichia coli*. Mutant proteins with alanine at position 66 replaced by glycine, eIF-4A(A66G), or valine, eIF-4A(A66V) were purified from *Escherichia coli* extracts and analyzed *in vitro* for activity in ATP crosslinking, ATP hydrolysis, RNA helicase and translation assays (Blum et al., 1992). The results show that *in vitro* ATP hydrolysis activity, RNA helicase activity and translation activity of eIF-4A correlate with *in vivo* activity of the factor. Whereas eIF-4A(A66G) showed wild-type activity in all assays, eIF-4(A66V) was active in ATP crosslinking but inactive in ATP hydrolysis and RNA helicase assays. *In vitro* translation was supported by wild-type eIF-4A and eIF-4A(A66G) but not by eIF-4A(A66V). The results show that the majority of mRNAs from *Saccharomyces cerevisiae* including an mRNA with the initiator AUG positioned eight nucleotides downstream of the cap structure require for their translation eIF 4A which is able to hydrolyze ATP

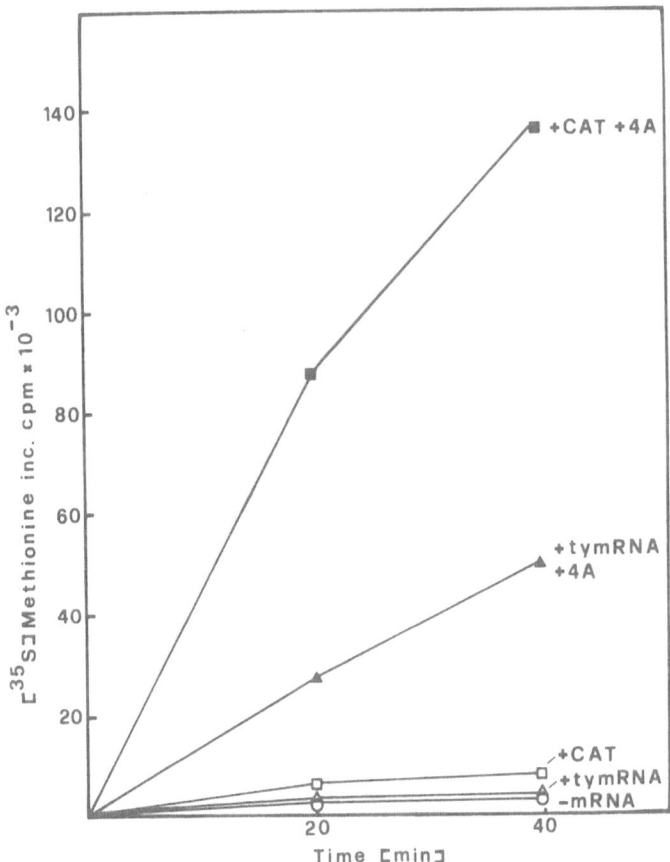

Fig. 2: Cell-free translation in eIF-4A-dependent extract. Initiation factor 4A-dependent extract was digested with micrococcal nuclease, reaction mixtures (20 µl) were incubated at 25°C, and 5 µl aliquots were analyzed for [35S]methionine incorporation. o, no addition; Δ, plus 10 µg of total yeast RNA (tymRNA); ❑, plus 50 ng chloramphenicol acetyl transferase (CAT) mRNA; ▲, plus 10 µg tymRNA plus 1 µg purified eIF-4A; ■, plus 50 ng CAT mRNA plus 1 µg purified eIF-4A.

# References

Altmann M, Handschin C, Trachsel H (1987) mRNA cap-binding protein: cloning of the gene encoding protein synthesis initiation factor eIF-4E from Saccharomyces cerevisiae. Mol Cell Biol 7(3):998-1003

Altmann M, Sonenberg N, Trachsel H (1989) Translation in Saccharomyces cerevisiae: initiation factor 4E- dependent cell-free system. Mol Cell Biol 9(10):4467-4472

Blum S, Mueller M, Schmid SR, Linder P, Trachsel H (1989) Translation in Saccharomyces cerevisiae: initiation factor 4A- dependent cell-free system. Proc Natl Acad Sci USA 86(16):6043-6046

Blum S, Schmid SR, Pause A, Buser P, Linder P, Sonenberg N, Trachsel H (1992) ATP Hydrolysis by Initiation Factor-4A Is Required for Translation Initiation in Saccharomyces-Cerevisiae. Proc Natl Acad Sci USA 89(16):7664-7668

Hanic-Joyce PJ, Singer RA, Johnston GC (1987) Molecular characterization of the yeast PRT1 gene in which mutations affect translation initiation and regulation of cell proliferaton. J Biol Chem 262:2845-2851

Keierleber C, Wittekind M, Qin S, McLaughlin CS (1986) Isolation and characterization of PRT1, a gene required for the initiation of protein biosynthesis in Saccharomyces cerevisiae. Mol Cell Biol 6:4419-4424

Linder P, Lasko PF, Leroy P, Nielsen PJ, Nishi K, Schnier J, Slonimsky PP (1989) Birth of the D-E-A-D box. Nature 337:121-122

# RNA - PROTEIN RECOGNITION AND BASIC MECHANISMS OF TRANSLATIONAL REPRESSION IN PROKARYOTES

D. E. Draper
Department of Chemistry
Johns Hopkins University
Baltimore, MD 21218
USA

## Introduction

The expression of many prokaryotic genes is regulated at the translational level by the binding of specific repressor proteins to mRNA. The first example of this kind of translational repression was the shut-off of RNA phage replicase translation by the phage coat protein (Nathans et al., 1969). Many other examples from phages have been described since, and translational repression contributes to the regulation of most of the ribosomal proteins (Nomura et al., 1984). Although the phenomenon has been known for 20 years, the mechanisms by which repression takes place are only now beginning to be understood. The problem of translational repression can be divided conveniently into two main questions. First, how do a protein and mRNA associate specifically? (I have avoided phrasing this as "how does a protein recognize an mRNA", to emphasize the participation of the RNA in this process.) Only in the last few years has atomic-resolution data on protein-RNA complexes become available, and it contains some surprises not anticipated from studies of protein-DNA complexes. The second question is how a protein bound to an RNA is able to inhibit the process of translation. Again, the first detailed studies of these mechanisms have given surprising answers to this question. The current understanding of these two questions will be reviewed here.

## Classes of RNA structures recognized by proteins

A large number of specific RNA binding proteins have been described, and it is worth asking whether there are any themes in the kinds of RNA structures recognized. Is there a minimum size, or minimum level of complexity, required to insure specificity? Do proteins "prefer" to make contacts with RNA bases, or with the RNA backbone? Several example proteins, most of them translational

NATO ASI Series, Vol. H 97
Post-transcriptional Control of Gene Expression
Edited by Orna Resnekov and Alexander von Gabain
© Springer-Verlag Berlin Heidelberg 1996

repressors, will illustrate the range of recognition mechanisms which are possible, starting with simple single stranded sequences and working up to RNA domains containing substantial tertiary structure.

The phage T4 regA gene represses the translation of a number of other T4 genes(Miller *et al.*, 1987). Mutations (Unnithan *et al.*, 1990), RNase protection (Winter *et al.*, 1987; Webster *et al.*, 1989), and measurements of oligomer-RNA binding affinities (Webster & Spicer, 1990) all demonstrate that the recognition site is contained within a 12 nucleotide sequence located within the ribosome binding site. The binding regions are not predicted to contain any strong secondary structures, and the recognition would appear to be a simple case of a protein binding a short, single-stranded sequence. An attempt to align the ribosome binding site regions of regulated mRNAs unexpectedly failed to reveal any consensus sequence (Webster *et al.*, 1989). Presuming that the RNA sequences must adopt some unusual structure which serves as a cue for specific binding, NMR studies were carried out on a 12mer specifically recognizing the regA protein, and on a second 12mer differing by one nucleotide and having 100-fold weaker affinity for the protein (Szewczak *et al.*, 1991). Both RNAs have unremarkable structures expected for single stranded nucleic acids with strong stacking between the bases. The basis for specific discrimination thus remains mysterious. A remaining possibility which has not yet been tested is that the recognized sequences can all adopt some peculiar structure or conformation once bound to the protein, and it is the ease with which this structure is adopted which determines the protein binding specificity.

Extensive studies on the R17 RNA phage coat protein of the replicase mRNA ribosome binding site have been carried out in Uhlenbeck's laboratory over a period of years (Uhlenbeck *et al.*, 1990). The minimal protein recognition site is a small hairpin and 5' tail, of 19 total nucleotides. Measurement of the binding affinities of nearly 100 sequence variants have shown that only a few sequence features are needed,as long as the hairpin framework is maintained. Among these are a pyrimidine and a purine in the hairpin loop; a Michael adduct formed between the protein and the 5,6 double bond of the pyrimidine probably takes place (Romaniuk & Uhlenbeck, 1985). A single base purine bulge within the helix is also required. Replacement of the purine with a variety of modified bases failed to identify any hydrogen bonding positions as essential for the recognition (Wu & Uhlenbeck, 1987). Either the ability of the purine to stack (e.g. with an aromatic amino acid side chain), or the distortion of the helix caused by a purine bulge, must be responsible for protein recognition of this base. A homologous coat protein from

Qβ phage also recognizes a small hairpin with a single base bulge, but in that case the bulge can be deleted without any adverse effect on the recognition.

Slightly more complex structures are recognized by several ribosomal proteins which are also translational repressors. Comparison of the rRNA and mRNA recognition sites reveals similarities at both the sequence and structural levels for these sites (Draper, 1989). S8 binds to a helical segment of about 15 base pairs, but there are several bulged As within this sequence which may stack together and significantly distort the axis and twist of the helix (Mougel *et al.*, 1987). The L1 recognition site is an internal loop between two short helices. Presumably it is a combination of the unique, irregular three dimensional structure of these sites, and the specific bases contained within them, which are used by the proteins to discriminate specific from non-specific sites.

The ribosomal RNA recognized by protein L11 represents a higher level of complexity. The minimal site is 74 nucleotides, though a 58 nucleotide fragment still has substantial binding specificity (Ryan & Draper, 1989; Draper, 1993b). The RNA is basically a junction of several helices, and thermal denaturation studies suggest that it contains considerable tertiary structure (Laing & Draper, 1993). Every base within the main 58 nucleotide recognition domain has been changed to its Watson-Crick complement, while preserving the phylogenetically conserved secondary structure (Ryan *et al.*, 1991). Only a few of these mutations decrease the L11 binding affinity by more than five fold, and these are all clustered about the junction region. The same mutations also affect the binding of an antibiotic, which suggests that the bases are not specifically contacted by the protein, but rather are needed to maintain the recognized tertiary structure. Preliminary melting experiments with these mutants confirms this suspicion (L. Laing and D. E. D., unpublished observations). These experiments rule out hydrogen bonding to the RNA bases as a major source of binding free energy, and suggest that the protein recognizes the shape of the backbone instead.

At least two of the ribosomal RNA binding proteins, S4 and L24, appear to organize entire rRNA domains of several hundred nucleotides (Draper, 1993b). Not all of these nucleotides can be essential for protein binding. The S4 protein is also a translational repressor, and its minimal mRNA target site is ~110 nucleotides (Tang & Draper, 1989). While the structure of this site is still complex, it is considerably simplified from the 460 nucleotides required to form the rRNA recognition site.

The best studied protein - RNA complexes may be the those of aminoacyl-tRNA synthetases with their cognate tRNAs. In a number of cases, mutations at a small number of sites convert a tRNA from a substrate for one synthetase to a substrate for another. In alanine-tRNA synthetase, the 2-amino group of the G in a wobble G-U pair turns out to be a major determinant of the specificity (Musier-Forsyth *et al.*, 1991), and nearby 2' hydroxyls also contribute to the binding free energy (Musier-Forsyth & Schimmel, 1992). While the synthetases show how proteins may evolve to recognize specific base features, it should be realized that the specificity problem confronting synthetases is unique. A synthetase must discriminate between a large number of tRNAs, all of which have the same basic three dimensional structure. For the most part, protein contacts with the backbone (i.e., *shape* recognition) will contribute to the protein's total binding affinity but not to its discriminatory capacity. In most other protein-RNA recognition problems, recognition of a specific, three dimensional configuration of the backbone does contribute to the binding specificity. Whether it has been evolutionarily "easier" or more advantageous for proteins to evolve base - specific or backbone (shape) - specific recognition remains to be seen.

These examples show that proteins can specifically recognize a wide range of RNA structures, from simple single strands, to irregular helices and loops, to complex tertiary structures. Thus there is no simple answer to what kinds of RNA sequences or structures might be involved in translational regulation- virtually any aspect of an mRNA may serve as a protein recognition site.

## RNA structural distortions induced by proteins

In the last few years the structures of several protein-RNA complexes have been solved by x-ray crystallography or NMR. These first atomic resolution glimpses of how proteins recognize RNA have revealed surprisingly large distortion of the RNA. The two crystal structures of complexes are both of aminoacyl-tRNA synthetases bound to their cognate tRNAs, tRNA[gln] (Rould *et al.*, 1989) and tRNA[asp] (Ruff *et al.*, 1991). In both cases the protein covers the surface of the tRNA on the "inside" of the L-shape, making contacts with the bases shown by mutation studies to be specificity determinants. The surprising observation is that the tRNA has been substantially distorted by the binding of the protein. (The assumption that the free tRNA is similar to tRNA[phe] must be made in the case of tRNA[gln], since the structure of the tRNA alone has not been solved. For tRNA[asp], structures of both the free and bound tRNAs have been determined by crystallography and can be compared directly.) At the terminus of the aminoacyl helix, the gln-tRNA synthetase has apparently disrupted the last Watson-Crick base pair of the helix, in order to bring the 3' terminus into register with the aminoacylation site of the protein. The RNA strand makes a sharp bend, and new hydrogen bonds are formed to stabilize this kink. At the anticodon end of the molecule, the hairpin loop has been substantially distorted to bring the anticodon bases into the protein, where specific contacts are made with the bases (Rould *et al.*, 1991). A similar distortion of the tRNA[asp] anticodon loop takes place; in superimposing the complexed and uncomplexed tRNA[asp] crystals structures, displacements up to 20 Å are observed (Ruff *et al.*, 1991).

A particularly simple model for protein - RNA recognition has been described by Frankel and coworkers, who deduced that a single arginine residue of the retroviral tat protein could specifically recognize the RNA target sequence (Calnan *et al.*, 1991). The site is a UCU bulge sequence, and its structures in the presence and absence of an arginine analog have been determined by NMR methods (Puglisi *et al.*, 1992). In the free RNA, the bulge bases are stacked into the helix. When arginine binds, the stacking of these bases is disrupted, most probably due to hydrogen bonding of one of the Us in the major groove of an adjacent helix, with an A-U pair. This rearrangement forms a small pocket neatly accommodating the arginine side chain and allowing interaction of the arginine guanidinium group with two backbone phosphates.

Although these three examples are a small sample of all the possible protein-RNA interactions, it is striking that each recognition mechanism involves substantial distortion of the RNA structure in ways which were not anticipated from mutation or chemical modification experiments. It is thus dangerous to assume that a "free" RNA structure is the same one bound by the protein, and this has at least two major implications for the interpretation of experimental data. First, in considering how tightly and specifically a protein binds a particular RNA, one must take into account how easily the RNA is "deformed" by the protein. Two RNAs which make identical contacts with a protein could possibly have very different binding affinities, if the secondary or tertiary structure of one required more energy to distort. Second, interpretation of popular "footprint" experiments becomes more difficult. Distortions of the RNA by the protein will lead to major changes in chemical reactivities, which may have nothing to do with the specific contacts actually made by the protein. The "footprint" is thus a superposition of protein contact sites and structural distortion sites, and there is no way the protection experiment alone can distinguish the two classes.

## Protein repressor effects on translational initiation

Once a protein has specifically recognized and bound an mRNA, there are two basic classes of mechanisms which the protein may use to inhibit translation. The most commonly thought of is simple competition: binding of the protein and the ribosome are mutually exclusive, and bound protein therefore prevents the ribosome from finding the initiation site. This has been termed the "displacement" mechanism (Draper, 1987). It does not necessarily require the repressor to physically cover the ribosome initiation site; binding of the protein elsewhere in the mRNA could conceivably stabilize an RNA structure which conceals the initiation region. The mechanism does require a substantial repressor - mRNA binding affinity. Free 30S ribosomes are present in *E. coli* at several micromolar concentrations [depending on growth rate, see (Kjeldgaard & Gausing, 1974)], and bind to mRNAs with affinities of $10^7$ - $10^8$ M$^{-1}$ (Draper, 1987). One can think of the product of the free concentration and binding affinity as a unitless "binding potential", in this case on the order of 100. To obtain 50% repression, the repressor binding potential must also be 100. Pools of ribosomal proteins which are translational repressors are present in *E. coli* at ~0.4 μM, and their mRNAs are normally repressed by about 50% (Draper, 1987), therefore their mRNA affinities must be on the order of $100/(0.4 \times 10^{-6})$ = 2.5 x $10^8$. This is about an order of magnitude higher than actually measured, which causes some doubt whether this class of repressors binds tightly enough to prevent translation.

Repressors need not make a head-to-head confrontation with ribosomes, but may inhibit translation by a more subtle strategy. A repressor may bind to the ribosome - mRNA complex and prevent some subsequent step of translational initiation, for instance binding of tRNA$_f^{met}$, or translocation after the first peptide bond is formed. This has been termed an "entrapment" mechanism (Draper, 1987). The repressor need not bind so tightly in this case, since it is, in effect, pinning the ribosome on the mRNA, rather than pushing it off. It is even possible that the repressor would bind the ribosome - mRNA complex more tightly than the mRNA alone; the protein might make some favorable contacts with the ribosome, or the ribosome might put the mRNA in a favorable conformation promoting repressor binding. There is at least one instance of a translational repressor, ribosomal protein S20, which does not show any specific affinity for its mRNA at all (Donly & Mackie, 1988).

There are very few instances in which a translational repression mechanism has been experimentally worked out, so it is not possible to say whether displacement or entrapment mechanisms are more commonly used. At least there are examples of each. As mentioned in the introduction, the coat protein of R17 phage is a translational repressor of the replicase gene. The RNA recognition site of the coat protein, a small hairpin, was discussed above. Both the Shine-Dalgarno and initiation codon sequences of the replicase gene are base paired within it. Therefore ribosomes and coat protein cannot simultaneously bind the replicase initiation site, and only the displacement model is feasible. The hairpin itself probably represses translation slightly, and the levels of coat protein which accumulate in vivo are probably sufficient to cause substantial additional repression (Draper, 1993a).

Recent experiments with the ribosomal protein S4, which regulates translation from the *E. coli* α operon, have shown that it uses a somewhat complex version of the entrapment mechanism. The messenger RNA binds 30S subunits in two different conformations: one able to bind $tRNA_f^{met}$ tightly in an initiation complex (the "active" complex), and the other incompetent in this activity ("inactive" complex) (Spedding *et al.*, 1993). S4 binds and stabilizes the inactive complex (Spedding & Draper, 1993). S4 is probably another case of a protein which binds and traps an RNA conformation which is not the most stable one.

There is also evidence that transcription repressors bind DNA simultaneously with the polymerase in some cases (Goodrich & McClure, 1992). Entrapment-type mechanisms may be a common way for repressor proteins to control gene expression, both at the transcriptional and translational levels.

## Acknowledgment

Work on RNA-protein interactions and translational regulation in the author's laboratory is supported by NIH grant GM29048.

## References

Calnan BJ, Tidor B, Biancalana S, Hudson D, Frankel AD (1991) Arginine-Mediated RNA Recognition: The Arginine Fork. Science 252:1167-1171

Donly BC, Mackie GA (1988) Affinities of ribosomal protein S20 and C-terminal deletion mutants for 16S rRNA and S20 mRNA. Nucleic Acids Res. 16:997-1010

Draper DE (1987) Translational Regulation of Ribosomal Proteins in *Escherichia coli*. Molecular Mechanisms. In:" Ilan J (eds) Translational Regulation of Gene Expression, Plenum Press, New York, pp. 1 - 26

Draper DE (1989) How do proteins recognize specific RNA sites? New clues from autogenously regulated ribosomal proteins. Trends Biochem. Sci. 14:335-338

Draper DE (1993) Mechanisms of Translational Initiation and Repression in Prokaryotes. In:" Nierhaus K (eds) The Translational Apparatus, Plenum, New York, in press

Draper DE (1993) Ribosomal Protein - RNA Interactions. In:" Dahlberg A Zimmermann R (eds) Ribosomal RNA: STructure, EVolution, Processing and Function in Protein Synthesis, CRC Press, Caldwell, NJ, in press

Goodrich JA, McClure WR (1992) Regulation of Open Complex Formation at the *Escherichia coli* Galactose Operon Promoters. Simultaneous Interaction of RNA Polymerase, *gal* Repressor and CAP/cAMP. J. Mol. Biol. 224:15-29

Kjeldgaard NO, Gausing K (1974) Regulation of Biosynthesis of Ribosomes. In:" Nomura M, Tiessières A Lengyel P (eds) Ribosomes, Cold Spring Harbor Laboratory, pp. 369-392

Laing LG, Draper DE (1993) Folding of a highly conserved fragment of the large subunit ribosomal RNA, in preparation

Miller ES, Karam J, Dawson M, Trojanowska M, Gauss P, Gold L (1987) Translational Repression: Biological Activity of Plasmid-encoded Bacteriophage T4 RegA Protein. J. Mol. Biol. 194:397-410

Mougel M, Eyermann F, Westhof E, Romby P, Expert-Bezançon A, Ebel J-P, Ehresmann B, Ehresmann C (1987) Binding of *E. coli* ribosomal protein S8 to 16S rRNA. A model for the interaction and the tertiary structure of the RNA binding site. J. Mol. Biol. 198:91-107

Musier-Forsyth K, Schimmel P (1992) Functional contacts of a transfer RNA synthetase with 2'-hydroxyl groups in the RNA minor groove. Nature 357:513-515

Musier-Forsyth K, Usman N, Scaringe S, Doudna J, Green R, Schimmel P (1991) Specificity for Aminoacylation of an RNA Helix: An Unpaired, Exocyclic Amino Group in the Minor Groove. Science 253:784-786

Nathans D, Oeschger MP, Polmar SK, Eggen K (1969) Regulation of Protein Synthesis Directed by Coliphage MS2 RNA I. Phage Protein and RNA Synthesis in Cells Infected with Suppressible Mutants. J. Mol. Biol. 39:279-292

Nomura M, Gourse R, Baughman G (1984) Regulation of the Synthesis of Ribosomes and Ribosomal Components. Ann. Rev. Biochem. 53:75-117

Puglisi JD, Tan R, Calnan BJ, Frankel AD, Williamson JR (1992) Conformation of the TAR RNA-arginine complex by NMR spectroscopy. Science 257:76-80

Romaniuk PJ, Uhlenbeck OC (1985) Nucleoside and Nucleotide Inactivation of R17 Coat Protein: Evidence for a Transient Covalent RNA-Protein Bond. Biochemistry 24:4239-4244

Rould MA, Perona JJ, Söll D, Steitz TA (1989) Structure of *E. coli* Glutamyl-tRNA Synthetase Complexed with tRNA$^{Gln}$ and ATP at 2.8 Å Resolution. Science 246:1135-1142

Rould MA, Perona JJ, Steitz TA (1991) Structural basis of anticodon loop recognition by glutaminyl-tRNA synthetase. Nature 352:213-218

Ruff M, Krishaswamy S, Boeglin M, Poterszman A, Mitschler A, Podjarny A, Rees B, Thierry JC, Moras D (1991) Class II Aminoacyl Transfer RNA Synthetases: Crystal Structure of Yeast Aspartyl-tRNA Synthetase Complexed with tRNA$^{Asp}$. Science 252:1682-1689

Ryan PC, Draper DE (1989) Thermodynamics of Protein-RNA Recognition in a Highly Conserved Region of the Large Subunit Ribosomal RNA. Biochemistry 28:9949-9956

Ryan PC, Lu M, Draper DE (1991) Recognition of the Highly Conserved GTPase Center of 23 S Ribosomal RNA by Ribosomal Protein L11 and the Antibiotic Thiostrepton. J. Mol. Biol. 221:1257-1268

Spedding GS, Draper DE (1993) Allosteric Mechanism for Translational Repression in the E. coli α Operon. Proc. Natl. Acad. Sci. USA, in press

Spedding GS, Gluick TC, Draper DE (1993) Ribosome Initiation Complex Formation with the Pseudoknotted α Operon Messenger RNA. J. Mol. Biol. 229: in press

Szewczak AA, Webster KR, Spicer EK, Moore PB (1991) An NMR Chracterization of the regA Protein-binding Site of Bacteriophage T4 Gene 44 mRNA. J. Biol. Chem. 266:17832-17837

Tang CK, Draper DE (1989) An Unusual mRNA Pseudoknot Structure is Recognized by a Protein Translational Repressor. Cell 57:531-536

Uhlenbeck OC, Gott JM, Witherell GW (1990) The Specific Interaction Between RNA Phage Coat Proteins and RNA. 40:185-220

Unnithan S, Green L, Morrissey L, Binkley J, Singer B, Karma J, Gold L (1990) Binding of the bacteriophage T4 regA protein to mRNA targets: an initiator AUG is required. Nucleic Acids Res. 18:7083-7092

Webster KR, Adari HY, Spicer EK (1989) Bacteriophage T4 RegA protein binds to the Shine-Dalgarno region of gene 44 mRNA. Nucleic Acids Res 17:10047-10068

Webster KR, Spicer EK (1990) Characterization of Bacteriophage T4 regA Protein-Nucleic Acid Interactions. J. Biol. Chem. 265:19007-19014

Winter RB, Morrissey L, Gauss P, Gold L, Hsu T, Karam J (1987) Bacteriophage T4 regA protein binds to mRNAs and prevents translation initiation. Proc. Natl. Acad. Sci. USA 84:7822-7826

Wu H-N, Uhlenbeck OC (1987) Role of a Bulged A Residue in a Specific RNA-Protein Interaction. Biochemistry 26:8221-8227

# TRANSLATIONAL CONTROL IN *Escherichia coli* AND *Saccharomyces cerevisiae*

John E.G. McCarthy
Gesellschaft für Biotechnologische Forschung mbH
Dept. of Gene Expression
Mascheroder Weg 1
D-3300 Braunschweig    FRG

## Introduction

In this short article I will briefly review some of the basic principles of translational control in the best-studied bacterium (*Escherichia coli*) and the best-studied lower eukaryote (*Saccharomyces cerevisiae*). Studies of these two organisms alone provide us with enormous insight into the mechanisms of translational control and regulation in prokaryotes and eukaryotes in general. In order to save space, I will frequently refer to review articles that cover specific areas, or at least cite a number of suitable publications of experimental work relevant to the themes discussed.

## Translational control in prokaryotes

A number of recent reviews have discussed various aspects of this theme (Gold, 1988; de Smit and van Duin, 1990; McCarthy and Gualerzi, 1990), and my intention is simply to provide an up-to-date summary based on these earlier contributions. It is still not clear how a given mRNA molecule determines the rate of its own translation in the cell. As discussed previously (McCarthy and Gualerzi, 1990), the major site of control and regulation of translation is the initiation step. Variations in the peptide elongation rate would be expected to influence the spacing between, and thus loading density of ribosomes, which in

NATO ASI Series, Vol. H 97
Post-transcriptional Control of Gene Expression
Edited by Orna Resnekov and Alexander von Gabain
© Springer-Verlag Berlin Heidelberg 1996

turn influences the overall availability of free ribosomes (pool size). However, the steady-state rate of production of complete peptides per unit time, and particularly the relative rate of completion of peptide chains from any given mRNA will be primarily determined by the initiation rate. But how might the sequence and intramolecular higher-order structure of an mRNA molecule determine the rate of initiation?

There are two types of recognition signal generally associated with translational initiation: the start codon (always present) is almost invariably preceded by a Shine-Dalgarno sequence (Fig. 1). Apart from these two sequence elements, there is no doubt that other sequences influence the efficiency of the initiation process. These are generally restricted to a region of the mRNA surrounding the start codon which can be defined in functional terms as the Translational Initiation Region (TIR; Fig. 1). These additional sequences can be of significance in two ways. First, it has been proposed by several groups that additional recognition elements exist which participate in mRNA-rRNA or mRNA-protein interactions involving the ribosome (usually the 30S subunit), thus influencing ("enhancing") the efficiency of start site localization (Table 1). In most cases very little, if any, direct evidence for the proposed interactions exists (see e.g. Firpo and Dahlberg, 1990). Only the existence of the ribosomal subunit S1: mRNA interaction seems to have been confirmed (by means of binding experiments; Boni et al., 1991). At the same time, it is difficult to assess the relative contributions of specific interactions with the additional sequence elements to the overall control of translation. One major reason for this is that their significance may be masked by the influence of higher-order structure in the TIR. The types and stabilities of the structures formed by intramolecular base-pairing [and other base interactions (stacking etc.)] will of course be determined by the sequence of the mRNA.This is the second way in which the sequence of the TIR can influence translational initiation. Intramole-

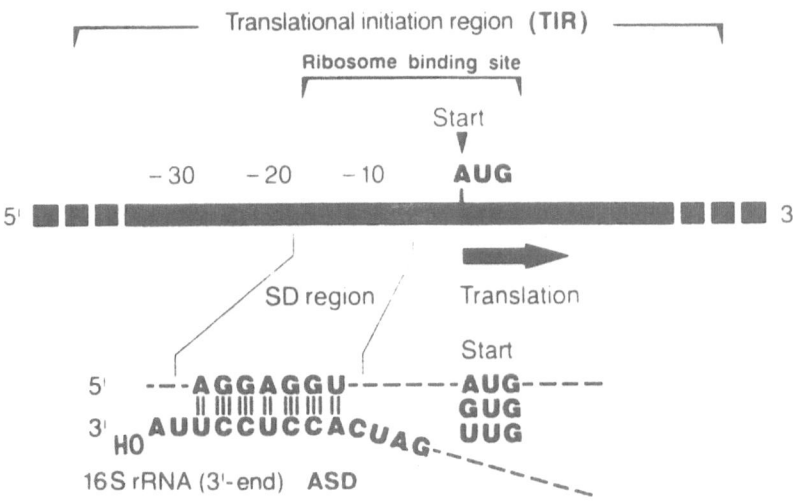

**Fig. 1 The prokaryotic (*E. coli*) ribosome-binding site (RBS) and the concept of the translational initiation region (TIR).** The Shine-Dalgarno (SD) region is complementary to bases proximal to the 3' -end of the 16S rRNA (the ASD region). The distance between the SD region and the start codon in recognised TIRs is 3-13 bases. The minimal definition of the 'classical' RBS comprises the SD region and the start codon plus the bases in between. The most common start codon is AUG (~ 90% of known *E. coli* genes), followed by GUG (~ 8%) and UUG (~ 1%), whereby initiation can also occur at AUU (see Table 1). At the other extreme, the RBS might be defined as the expanse of mRNA protected from ribonuclease digestion by bound ribosomes. The 5' part of the TIR may comprise an untranslated leader or alternatively may include the end of an upstream gene (on a polycistronic message), and thus a translational stop codon. The primary sequence of the TIR contains not only recognition elements but also encodes 'blue-prints' for possible modes of higher order structure. If the TIR (as opposed to the RBS) is to be defined as that region determining both the site and the efficiency of initiation, the definition will include bases beyond the 5' and 3' limits of the RBS.

**TABLE 1  CONFIRMED/PROPOSED RECOGNITION ELEMENTS IN THE *E.coli* TIR**

| SEQUENCE ELEMENT | POSITION RELATIVE TO START CODON | (PROPOSED) FUNCTION | EVIDENCE/COMMENTS |
|---|---|---|---|
| **Confirmed recognition sequences** | | | |
| Start codons (AUG, GUG, UUG) | +1 to +3 | Initiation via interaction with fMet-tRNA$^{fMet}$ | Usually only possible choices (but see AUU in *infC*) [1] |
| Shine-Dalgarno region: 3 or more of UAAGGAGGUGA | Usually within the region -15 to -2 | Interaction with 3' end of 16S rRNA (bases 1532-1542, especially 1535-1539) | Clearest evidence: 'specialised' ribosome systems [2, 3] |
| **Proposed (additional) recognition elements** | | | |
| Preference for specific bases, especially As | Spacer region, 5' of SD, 3' of start | See 'enhancing' sequences (below) | Sequence analyses/expression with manipulated TIRs [4-9] |
| Complementarity to bases adjac. to ASD: any 3 or more of UGAUCC | Within region -58 to -4 | Extra recognition element compl. to 16S rRNA (bases 1529-1534) 'initiation promoting site' | Sequence analyses/assumed initiation efficiencies [10] |
| At least 3 of: UCAAAACUCUUCAAUUU | Within region +4 to +25 | Extra recognition element compl. to 16S rRNA (bases 1-18) | Sequence analyses [11] |
| 'Omega' sequence - core is: ACAAUUAC | Region not clearly defined | Extra recognition element active in various organisms | Data obtained using sequence from tobacco mosaic virus [12] |

| | | | |
|---|---|---|---|
| Extra *infC* interaction sites: AAGGUAUUAA and GGCGG | -5 to +5 and +7 to +11, resp. | Interaction with 16S rRNA (bases 463-472 and 1399-1403) | Importance of AUU start for IF3 feedback confirmed [1, 13] |
| Extended complementarity to tRNA^fMet: GUUAUGAGC | -3 to -1 and +4 to +6 | More stable interaction with tRNA^fMet a-loop | Sequence analysis [14] |
| 'Downstream box' with up to 12 complementary bases | Various positions (gene-dependent) | Additional interaction with bases 1469 to 1483 of 16S rRNA | Deletion analysis [21-24] |
| Other 'enhancing' sequences (e.g. in TIRs of *atpE*, bacteriophage T7 gene 10) | Within region approx. -50 to +50 | At least to some extent due to highly accessible SD/start region (open TIR); see also [20] | Data obtained using various natural and manipulated TIRs [15, 16] |
| S1 binding site: polyU | In vicinity of SD | Ribosomal subunit S1 binding | Analyses of interactions with mRNA [17-19] |

References:

[1] Gold, 1988
[2] Hui and de Boer, 1987
[3] Jacob et al., 1987
[4] Dreyfus, 1988
[5] Stormo, 1986
[6] Gold et al., 1981
[7] Scherer et al., 1980
[8] Kozak, 1983
[9] Hui et al., 1984
[10] Thanaraj and Pandit, 1989
[11] Petersen et al., 1988
[12] Gallie and Kado, 1989
[13] Gold et al., 1984
[14] Ganoza et al., 1985
[15] McCarthy et al., 1985
[16] Olins and Rangwala, 1989
[17] Boni et al., 1991
[18] McCarthy, 1987
[19] Subramanian, 1983
[20] Firpo and Dahlberg, 1990
[21] Sprengart et al., 1990
[22] Faxen et al., 1991
[23] Nagai et al., 1991
[24] Shean and Gottesmann, 1992

cular base-pairing within or involving the TIR can inhibit translational initiation to varying degrees (see Schauder and McCarthy, 1989; McCarthy and Gualerzi, 1990; de Smit and van Duin, 1990, and the references in these papers). A further factor that needs to be considered in relation to the potential function(s) of TIR sequences that belong to the reading frame is that constraints are set not only by the choice of codons corresponding to the protein sequence but also probably by other requirements, for example dictated by frame monitoring mechanisms (see e.g. Lagunez-Otero and Trifonov, 1992).

How can we envisage the sequence and structure of the TIR acting to control initiation at the molecular level? The concept of "thermodynamic control" being exercised by mRNA structure has been discussed in general terms before (Stormo, 1986; McCarthy and Gualerzi, 1990). The most precise investigation of the relationship between the free energy of folding of a stem-loop structure and its effect upon translational efficiency has been performed using the bacteriophage MS2 coat protein TIR (de Smit and van Duin, 1990). The results of this study are fully consistent with a mechanism of control in which a stem-loop structure renders the initiation site inaccessible to 30S ribosomal subunits. The thermodynamic stability of the stem-loop determines the fraction of the coat protein mRNA molecules that is accessible to ribosomal binding, and thus the degree of inhibition, although it should be remembered that the proportion of the mRNA molecules in the "open" or partially "open" state will also be influenced by other factors (such as protein binding). At the same time, this form of control does not exclude effects of TIR structure on the kinetics of the individual steps of the initiation pathway. One model of "kinetic control", based on the McClure model of promoter function, has been discussed by Ringquist et al. (1992). Again, it remains uncertain to what extent additional "recognition signals" in the TIR (Table 1), as opposed to mRNA structure, contribute to either of the types of control mentioned here.

The sequences and structures of the TIR could therefore influence the accessibility of the initiation site, or the kinetics of the initiation pathway, or both. The only way to differentiate between these effects is on the basis of precise kinetic data for individual steps of the initiation pathway and the results of studies of the binding interactions between ribosomes and mRNA, all of these data being obtained using a number of mRNA species, each one differing in the effective efficiency with which it is translated. Up to now, few laboratories have tried to perform experiments along these lines (see e.g. Gualerzi et al., 1988; Lang et al., 1989; Ringquist et al., 1992; Dontsova et al., 1992; de Smit and van Duin, 1990; and references therein), and a great deal more effort in this direction remains necessary. In this context, it is essential not to underestimate the number of possible types of structure that may be formed by mRNA. The potential for formation of not only Watson-Crick hydrogen-bonds but also non-orthodox base-pairs and stacking interactions increases the range of structural forms that can be assumed by this macromolecule (see e.g. Cheong et al., 1990; Puglisi et al., 1988; Heus and Pardi, 1991; Szewczak et al., 1991).

**Translational coupling in prokaryotes**

This widespread phenomenon has received far too little attention. Given that most bacterial genes are likely to be organized in operons and are frequently coupled at the translational level, we are surprisingly ignorant of the possible mechanisms that are involved. Translational coupling renders the function of a TIR dependent upon the translation of an upstream reading frame. Direct reinitiation by a ribosome that has terminated at a stop codon in close proximity to a given gene's start codon can allow for tight coupling with a maximum stoichiometry of translation of 1:1 (Adhin and van Duin, 1990; Hellmuth et al., 1991). In an alternative type of mechanism, a stem-loop structure in the TIR of a coupled gene prevents translation in the absence of translation of the upstream reading-frame. The coupling as such takes the form of a regulation of the accessibility of the initiation site to ribosomes (Kastelein et al., 1983; Hellmuth et al., 1991). However, little is known about the way(s) in which ribosomes terminating on the upstream reading frame of a coupled pair of genes facilitate the translation of the coupled downstream reading frame. Although "fresh" ribosomes from the cellular pool undoubtedly translate the downstream gene of many coupled gene pairs, the initial "opening" of the inhibitory secondary structure may necessarily involve a reinitiating ribosome. A further unclarified aspect of prokaryotic translational coupling is that of the kinetics of "melting" and (re)formation of the stem-loop structure. New data now indicate that in the case of the coupled pair *atpHA* the original stem-loop in the freshly transcribed *atpA* TIR acts as a sort of "gating device" that is "switched" to form a less inhibitory or non-inhibitory structure that does not rapidly refold to the initial structure (Rex et al., 1993).

## Translational control in yeast

Translational control in *S. cerevisiae* seems to follow the same general principles as those observed by all eukaryotic systems but certain differences to for example, animal cells, have become apparent. Recent reviews have summarized much of the literature on this subject (Raué et al., 1990; Linder and Prat, 1990; Hinnebusch and Liebman, 1991; Yoon and Donahue, 1992) and I want to concentrate only on one specific aspect of translational control in yeast: the role of mRNA secondary structure. Studies of the influence of secondary (and perhaps tertiary structure) in mRNA upon translation provide us with important basic information that is relevant to the mechanism of initiation. As we have seen, stem-loop structures play a central role in determining the relative translation rates in bacteria. They are also known to be able to inhibit translation in eukaryotes (Pelletier and Sonenberg, 1988; Kozak, 1986, 1989; Baim and Sherman, 1988; Cigan et al., 1988), but in the latter case much less is known about their functional significance. Kozak (1989, 1990) explains the inhibitory effect of stem-loop structures in terms of the "scanning" model of translational initiation. A stem-loop is proposed either to interfere with the binding of the initiation complex at the 5' end of the leader, or to block the "scanning" of this complex along the leader before it locates a start codon. It should be stressed that the effects of secondary structure upon the translation of specific mRNAs must be seen in the context of competition between a large number of different mRNAs in the cell (see e.g. the theoretical considerations of Godefroy-Colburn and Thach, 1981).

Translation in yeast is more sensitive to the presence of stem-loop structures than it is in higher eukaryotes (Baim and Sherman, 1988; Cigan et al., 1988; Vega Laso et al., 1993), while being generally less readily inhibited than bacterial translation. An important question is whether only the stability of a stem-loop in a eukaryotic leader is decisive in terms of the (quantitative degree of) inhibition of translation, or does the position relative to the 5' end and start codon of the mRNA also play a role? This question is generally easier to analyze in eukaryotes partly because more stable, and thus better charac-terized, stem-loops can be introduced into defined positions in the mRNA without completely inhibiting translation. Experiments performed using syn-thetic mRNAs in reticulocyte lysates have indicated that eukaryotic ribosomes are more effectively blocked by a stem-loop of given stability if this is positio-ned at the 5'-end of an mRNA rather than at the start codon (Kozak,1989). Results obtained using in vivo and in vitro systems of *S. cerevisiae* on the other hand, do not confirm this position effect, at least specifically for this lower eukaryote (Vega Laso et al., 1993; Oliveira et al., 1993). The positioning of a stem-loop at the 5'-end of a yeast leader does not heighten its effectivity. Indeed, it is likely to be a more potent inhibitory element when placed near, or when "straddling", the start codon of the translated reading frame. By contrast, a stem-loop structure can positively influence the selection of a specific AUG in a mammalian leader when placed immediately behind this start codon, possibly by "stalling" ribosomes and thus improving the chances of initiation at this point (Kozak, 1990). An interesting parallel has been described in gene 38 of bacteriophage T4, where a stem-loop brings a Shine-Dalgarno region that lies 23 nucleotides upstream of a gene's start codon into a position relative to this start codon that promotes initiation (Gold, 1988). Once again, we should also be aware of the possibility that not only stem-loop structures but also other types of intramolecular interactions can influence leader function. In this respect, it is of interest to examine the effects of base composition in yeast leaders upon the translational efficiency of the *PGK* gene (van den Heuvel et al., 1990).

Why are there such discrepancies between the inhibitory influence of secondary structures observed in yeast and mammalian cells (Table 2)? They may indicate the existence of differences in the mechanistic pathway of initiation between higher and lower eukaryotes, and it would therefore be worthwhile to determine what these differences are. On the other hand, perhaps the story is more complicated, in that not all higher eukaryotic cell types, or even tissue cell types of a given organism, behave identically in this respect. The initiation pathway, and thus possibly its sensitivity to the stability and position of stem-loop structures, may be modified by changes in the nature and/or relative amounts of the components (e.g. initiation factors) that are involved (see e.g. Kozak, 1990; Koromilas et al., 1992). As to whether the inhibitory effects of secondary structure on translation in eukaryotes in general, and in yeast in particular, can be fully explained by the "scanning" model remains to be established. The evidence that internal initiation can occur, at least in higher eukaryotes (see e.g. Pelletier and Sonenberg, 1988; Jackson et al., 1990; Mace-

**TABLE 2   INHIBITION OF TRANSLATION BY SECONDARY STRUCTURE**

| TYPE OF ORGANISM | FREE ENERGY OF FORMATION OF SECONDARY STRUCTURE NECESSARY FOR > 90% INHIBITION* | POSITION EFFECTS REPORTED | REFERENCES |
|---|---|---|---|
| Higher eukaryote (mammalian cell lines/ in vitro systems) | approx. -50 kcal mol$^{-1}$ | Stem-loop more effective at 5' end than close to initiation codon | Kozak, 1986; 1989 |
| S. cerevisiae | between -14 and -20 kcal mol$^{-1}$ | Stem-loop not more effective at 5' end. At least in some cases more inhibitory in region of start codon | Baim and Sherman, 1988 Cigan et al., 1988 Vega Laso et al., 1993 Oliveira et al., 1993 |
| E. coli | approx. -10 kcal mol$^{-1}$ | Less easily defined, but structure must interfere with direct accessibility of SD region, and possibly of start codon | Schauder and McCarthy, 1989 de Smit and van Duin, 1990 |

* These estimates of stability can only be approximate because of position effects.

jak and Sarnow, 1991) points to the existence of "alternative" pathways which, in one form or another, may also be present in yeast. How dissimilar to the "scanning pathway", and how similar to the bacterial type of initiation pathway, are these "alternative" pathways? This, and other questions touched upon in this short article, will only be answerable once more is understood of the molecular details of the protein-RNA and RNA-RNA interactions involved in translation. "With a rebel yell, more, more, more" (Idol, 83).

## References

Adhin MR, van Duin J (1990) Scanning model for translational reinitiation in eubacteria. J Mol Biol 213: 811-818

Baim SB, Sherman F (1988) mRNA structures influencing translation in the yeast *S. cerevisiae*. Mol Cell Biol 8: 1591-1601

Boni IV, Isaera DM, Musychenko ML, Tzareva NV (1991) Ribosome-messenger recognition: mRNA target sites for ribosomal protein S1. Nucleic Acids Res 19: 155-162

Cheong C, Varani G, Tinoco I Jr (1990) Solution structure of an unusually stable RNA hairpin, 5'GGAC(UUCG)GUCC. Nature 346: 680-682

Cigan AM, Feng L, Donahue TF (1988) tRNAi$^{met}$ functions in directing the scanning ribosome to the start site of translation. Science 242: 93-97

de Smit M, van Duin J (1990) Control of prokaryotic translational initiation by mRNA secondary structure. Progr Nucl Acid Res Mol Biol 38: 1-35

Dontsova O, Dokudovskaya S, Kopylov A, Bogdanov A, Rinke-Appel J, Jünke N, Brimacombe R (1992) Three widely separated positions in the 16S RNA lie in or close to the ribosomal decoding region; a site-directed cross-linking study with mRNA analogues. EMBO J 11: 3105-3116

Dreyfus M (1988) What constitutes the signal for the initiation of protein synthesis on *Escherichia coli* mRNAs? J Mol Biol 204: 79-94

Faxén M, Plumbridge J, Isaksson LA (1991) Codon choice and potential complementarity between mRNA downstream of the initiation codon and bases 1471 - 1480 in 16S ribosomal RNA affects expression of *glnS*. Nucleic Acid Res 19: 5247-5251

Firpo MA, Dahlberg AE (1990) The role of ribosomal RNA in the control of gene expression. In "Posttranscriptional Control of Gene Expression" (JEG McCarthy, M.F. Tuite, eds.), Springer-Verlag, Berlin, pp. 185-195

Gallie DR, Kado CI (1989) A translational enhancer derived from tobacco mosaic virus is functionally equivalent to a Shine-Dalgarno sequence. Proc Natl Acad Sci USA 86: 129-132

Ganoza M, Marliere P, Kofoid E, Louis BG (1985) Initiator tRNA may recognise more than the initiation codon in mRNA: a model for translational initiation. Proc Natl Acad Sci USA 82: 4587-4591

Godefroy-Colburn T, Thach RE (1981) The role of mRNA competition in regulating translation. J Biol Chem 256: 11762-11773

Gold L (1988) Posttranscriptional regulatory mechanisms in *Escherichia coli*. Ann Rev Biochem 57: 199-233

Gold L, Pribnow D, Schneider T, Shinedling S, Singer BS, Stormo G (1981) Translational initiation in prokaryotes. Ann Rev Microbiol 35: 365-403

Gold L, Stormo G, Saunders R (1984) *Escherichia coli* translational initiation factor IF3: a unique case of translational regulation. Proc Natl Acad Sci USA 81: 7061-7065

Gualerzi CO, Calogero RA, Canonaco MA, Brombach M, Pon CL (1988) Selection of mRNA by ribosomes during prokaryotic translational initiation. In "Genetics of Translation" (MF Tuite, M Picard, M Bolotin-Fukuhara, eds.) Springer-Verlag, Berlin, 317-330

Hellmuth H, Rex G, Surin B, Zinck R, McCarthy JEG (1991) Translational coupling varying in efficiency between different pairs of genes in the central region of the *atp* operon of *Escherichia coli*. Molec Microbiol 5: 813-824

Heus HA, Pardi A (1991) Structural features that give rise to the unusual stability of RNA hairpins containing GNRA loops. Science 253: 191-194

Hinnebusch AG, Liebman SW (1991) Protein synthesis and translational control in *Saccharomyces cerevisiae*. In "The Molecular and Cellular Biology of the yeast *Saccharomyces*" (JR Broach, JR Pringle, EW Jones, eds.) Cold Spring Harbour Press, NY, USA, pp. 627-735

Hui A, de Boer H (1987) Specialised ribosome system: preferential translation of a single mRNA species by a subpopulation of mutated ribosomes in *Escherichia coli*. Proc Natl Acad Sci USA 84: 4762-4766

Hui A, Hayflick J, Dinkelspiel K, de Boer HA (1984) Mutagenesis of the three bases preceding the start codon of the ß-galactosidase mRNA and its effect on translation in *Escherichia coli*. EMBO J 3: 623-629

Jackson RJ, Howell MT, Kaminski A (1990) The novel mechanism of initiation of picornavirus RNA translation. Trends Biochem Sci 15: 477-483

Jacob WF, Santer M, Dahlberg AE (1987) A single base change in the Shine-Dalgarno region of 16S rRNA of *Escherichia coli* affects translation of many proteins. Proc Natl Acad Sci USA 84: 4757-4761

Kastelein RA, Berkhout B, van Duin J (1983) Opening the closed ribosome-binding site of the lysis cistron of bacteriophage MS2. Nature 305: 741-743

Koromilas AE, Lazaris-Karatzas A, Sonenberg N (1992) mRNAs containing extensive secondary structure in their 5' non-coding region translate efficiently in cells overexpressing initiation factor eIF-4E. EMBO J 11: 4153-4158

Kozak M (1983) Comparison of initiation of protein synthesis in procaryotes, eucaryotes and organelles. Microbiol Rev 47: 1-45

Kozak M (1986) Influences of mRNA secondary structure on initiation by eukaryotic ribosomes. Proc Natl Acad Sci USA 83: 2850-2854

Kozak M (1989) Circumstances and mechanisms of inhibition of translation by secondary structure in eukaryotic mRNAs. Mol Cell Biol 9: 5134-5142

Kozak M (1990) Structural features in eukaryotic mRNAs that modulate the initiation of translation. J Biol Chem 266: 19867-19870

Lagunez-Otero J, Trifonov EN (1992) mRNA Periodical infrastructure complementary to the proof-reading site in the ribosome. J Biomolec Struct & Dynam 10: 455-464

Lang V, Gualerzi C, McCarthy JEG (1989) Ribosomal affinity and translational initiation in *Escherichia coli*. J Mol Biol 210: 659-663

Linder P, Prat A (1990) Baker's yeast, the new work horse in protein synthesis studies: analyzing eukaryotic translation initiation. Bio Essays 12: 519-526

McCarthy JEG (1987) Expression of the unc genes in *Escherichia coli*. J Bioenerg Biomemb 20: 19-39

McCarthy JEG, Gualerzi C (1990) Translational control of prokaryotic gene expression. Trends in Genetics 6: 78-85

McCarthy JEG, Schairer HU, Sebald W (1985) Translational initiation frequency of *atp* genes from *Escherichia coli*: identification of an intercistronic sequence that enhances translation. EMBO J 4: 519-526

Nagai H, Yuzawa H, Yura T (1991) Interplay of two cis-acting mRNA regions in translational control of $\delta^{32}$ synthesis during the heat shock response of *Escherichia coli*. Proc. Natl. Acad. Sci 88: 10515-10519

Olins PO, Rangwala SH (1989) A novel sequence element derived from bacteriophage T7 mRNA acts as an enhancer of translation of the *lacZ* gene in *Escherichia coli*. J Biol Chem 264: 16973-16976

Oliveira CC, van den Heuvel JJ, McCarthy JEG (1993) Inhibition of translational initiation in *Saccharomyces cerevisiae* by secondary structure: the roles of the stability and position of stem-loops in the mRNA leader. Molec Microbiol, in press

Pelletier J, Sonenberg N (1988) Internal initiation of translation of eukaryotic mRNA directed by a sequence derived from poliovirus RNA. Nature 334: 320-325

Petersen GB, Stockwell PA, Hill DF (1988) Messenger RNA recognition in Escherichia coli: a possible second site of interaction with 16S ribosomal RNA. EMBO J 7: 3957-3962

Puglisi JD, Wyatt JR, Tinoco I Jr (1988) A pseudoknotted RNA oligonucleotide. Nature 331: 283-286

Raué HA, van den Heuvel JJ, Planta RJ (1990) In "Posttranscriptional Control of Gene Expression" (JEG McCarthy and MF Tuite, eds.), Springer-Verlag, Berlin, pp. 237-247

Rex G, Surin B, Besse G, Schneppe B, McCarthy JEG (1993) The mechanism of translational coupling in Escherichia coli; higher-order structure in the atpHA mRNA acts as a "gating device" for ribosomes. In preparation

Ringquist S, Shinedling S, Barrick D, Green L, Binkley J, Stormo GD, Gold L (1992) Translation initiation in Escherichia coli: sequences within the ribosome-binding site. Molec Microbiol 6:1219-1229

Schauder B, McCarthy JEG (1989) The role of bases upstream of the Shine-Dalgarno region and in the coding sequence in the control of gene expression in Escherichia coli: translation and stability of mRNAs in vivo. Gene 78: 59-72

Scherer GFE, Walkinshaw MD, Arnott S, Morre DJ (1980) The ribosome binding sites recognised by E.coli ribosomes have regions with signal character in both the leader and protein coding segments. Nucleic Acids Res 8: 3895-3907

Shean CS, Gottesman ME (1992) Translation of the prophage λ cl transcript. Cell 70: 513-522

Sprengart ML, Fatscher HP, Fuchs E (1990) The initiation of translation in E. coli: apparent base pairing between the 16srRNA and downstream sequences of the mRNA. Nucleic Acid Res 18: 1719-1723

Stormo GD (1986) Translation initiation. In: Maximizing Gene Expression, Reznikoff W, Gold L (eds), Butterworths, Boston, p195

Subramanian AR (1983) Structure and functions of ribosomal protein S1. Progr Nucleic Acid Res Mol Biol 28: 101-142

Szewczak AA, Webster KR, Spicer EK, Moore PB (1991) An NMR characterization of the regA protein-binding site of bacteriophage T4 gene 44 mRNA. J Biol Chem 266: 17832-17837

Thanaraj TA, Pandit MW (1989) An additional ribosome-binding site on mRNA of highly expressed genes and a bifunctional site on the colicin fragment of 16S rRNA from Escherichia coli: important determinants of the efficiency of translation-initiation. Nucleic Acids Res 17: 2973-2985

van den Heuvel JJ, Planta RJ and Raué HA (1990) Effect of leader primary structure on the translational efficiency of phosphoglycerate kinase mRNA in yeast. Yeast 6: 473-482

Vega Laso MR, Zhu D, Sagliocco F, Brown AJP, Tuite MF, McCarthy JEG (1993) Inhibition of translational initiation in the yeast *Saccharomyces cerevisiae* as a function of the stability and position of hairpin structures in the mRNA leader. J Biol Chem 268

Yoon H, Donahue TF (1992) Control of translation initiation in *Saccharomyces cerevisiae*. Molec Micriobiol 6: 1413-1419

# Translational Control in Eukaryotic Cells: Principles Learned from the IRE/IRF System

Matthias W. Hentze
European Molecular Biology Laboratory
Meyerhofstrasse 1
Postfach 10.2209
D-6900 Heidelberg 1
Federal Republic of Germany
Tel.: (-49)-6221-387 501
FAX: (-49)-6221-387 518

In addition to gene regulation within the nucleus, the expression of genetic information can also be controlled in the cytoplasm of eukaryotic cells. One of these cytoplasmic mechanisms is translational regulation, where a signal induces a change in the rate of translation of a constant amount of mRNA.

Iron regulation of ferritin expression has served as a model system to investigate translational regulation in mammalian cells (Zähringer et al., 1976; for reviews: Kühn and Hentze, 1992; Theil, 1990; Klausner and Harford, 1989). Apoferritin is an intracellular iron storage protein that consists of 24 assembled subunits of ferritin light (L) and heavy (H) chains. Both subunits originate from separate genes. Iron administration to rats or to tissue culture cell lines increased the synthesis of ferritin H- and L-chains in the absence of corresponding changes in ferritin mRNA levels. Conversely, administration of the iron chelator desferrioxamine reduced ferritin synthesis in live animals and tissue culture cells (Rogers and Munro, 1987; Hentze et al., 1987; Goossen et al., 1990). It was proposed that cellular iron levels regulated the translation of ferritin H- and L-chain mRNAs by a common mechanism. As will be described, this suggestion was found to be correct and in the following pages all references to "ferritin" are

NATO ASI Series, Vol. H 97
Post-transcriptional Control of Gene Expression
Edited by Orna Resnekov and Alexander von Gabain
© Springer-Verlag Berlin Heidelberg 1996

intended to include both chains of the protein or their respective mRNAs.

Ferritin mRNAs contain a ~30 nucleotide element in their 5' untranslated regions (UTRs) which is necessary and sufficient to mediate translational iron regulation. This element has been named "iron-responsive element = IRE" (Hentze et al., 1987; Aziz and Munro, 1987; Caughman et al., 1988). The IRE represents a characteristic and phylogenetically highly conserved stem and loop structure with the following features: [a] a six nucleotide loop with the sequence 5' CAGUGN 3' (N= C, U or A); [b] a top helix of five (or occasionally four paired nucleotides; [c] an unpaired 5' C residue separated from the loop by five nucleotides; and [d] a bottom helix of somewhat variable length (Hentze et al., 1988). In addition to its typical structure, the position of the IRE within the 5' UTR of ferritin mRNAs is conserved. In all IRE-containing ferritin mRNAs that have been characterized to date, the IRE is located 30-40 nucleotides downstream from the 5' cap structure. It was shown that the position of the IRE is important for its function, since its relocation to a position further downstream resulted in a strong reduction of its ability to mediate iron regulation (Goossen et al., 1990; Goossen and Hentze, 1992).

A cytoplasmic protein of ~98 kDa specifically binds with high affinity to IREs (Leibold and Munro, 1988; Hentze et al., 1989; Haile et al., 1989). This protein is referred to as "iron-regulatory factor (IRF)", "IRE-binding protein (IRE-BP)", "ferritin repressor protein (FRP)", or "P90". As implied by one of its names (ferritin repressor protein), binding of IRF to the IRE in ferritin mRNAs causes repression of translation both *in vivo* and *in vitro* (Brown et al., 1989; Goossen et al., 1990). In addition to this function, IRF also regulates the translation of erythroid 5-aminolevulinate synthase (eALAS) mRNA by binding to a cap-proximal IRE (Dandekar et al., 1991; Cox et al., 1991; Melefors et al., 1993) and the stability of transferrin receptor (TfR) mRNA *via* a region in the 3'UTR of TfR mRNA that includes 5 IREs (Casey et al., 1988; Müllner et al., 1989). Thus, IRE/IRF interactions co-ordinate the post-transcriptional control of the expression of proteins involved in iron uptake (transferrin receptor), iron storage (ferritin) and iron utilization (eALAS as the first enzyme in the erythroid heme biosynthetic pathway). Additional roles for IRE/IRF are suggested by the occurrence of an IRE in the 5' UTR of aconitase mRNA (Dandekar et al., 1991).

Iron regulation of ferritin and eALAS translation as well as TfR mRNA stability results from iron-dependent control of the IRE-binding activity of IRF. In iron-starved cells, high affinity binding of IRF to IREs is stimulated, resulting in repression of ferritin and eALAS mRNA translation and stabilization of TfR mRNA. In contrast, iron administration causes reduced IRE-binding of IRF, thus de-repressing translation and permitting rapid TfR mRNA decay (Leibold et al., 1988; Haile et al., 1989).

Several lines of evidence suggest that the regulation of IRF by iron occurs post-translationally. First, iron perturbations do not induce changes in IRF mRNA levels (Patino and Walden, 1992). Second, changes in the IRE-binding activity of IRF can be observed even in cells that have been pre-treated with puromycin, an inhibitor of protein synthesis (Hentze et al., 1989). Third, the low IRE-binding activity in extracts from iron-treated cells can be activated *in vitro* to levels comparable to the high IRE-binding activity present in extracts from iron-starved cells (Hentze et al., 1989; Haile et al., 1989).

To explain the post-translational mechanism by which IRF senses cellular iron levels and adjusts its IRE-binding activity, the "iron switch model" has been proposed. It is mainly based on the following findings: IRF is highly similar to the iron sulfur proteins aconitase and isopropylmalate isomerase (Hentze and Argos, 1991; Rouault et al., 1991). This similarity includes the precise positioning of three cysteines which participate in the co-ordination of cubane iron-sulfur clusters. Furthermore, IRF exhibits aconitase activity and thus likely belongs to the family of aconitase-like iron sulfur proteins (Kaptain et al., 1992; Haile et al., 1992; Quick et al., submitted). Purification of mitochondrial aconitase often results in loss of one iron atom from the cluster and concomitant loss of enzymatic activity. When this fourth iron is re-introduced *in vitro*, enzymatic activity is restored (Beinert and Kennedy, 1989).

Treatment of purified IRF or recombinant IRF with iron salts revealed that IRF is an iron-dependent bi-functional protein, capable to exert IRE-binding and aconitase activities (Kaptain et al., 1991; Constable et al., 1992; Haile et al., 1992; Quick et al., submitted). Addition of $FeSO_4$ (in a reducing buffer) to recombinant IRF expressed in *E.coli* causes loss of its IRE-binding and activation of its aconitase activity (Quick et al., submitted). These results suggest that the "aconitase form" of IRF possesses a 4Fe-4S cluster, whereas the "IRE-binding form" exists as a protein with a smaller cluster or as an iron-free apoprotein. Accordingly, iron regulation of IRF occurs post-translationally and involves interconversions of its iron sulfur cluster.

The repression of ferritin and eALAS mRNA translation by IRE/IRF complex formation has been reconstituted *in vitro* (Brown et al., 1989; Gray et al., submitted). In cell-free translation extracts prepared from rabbit reticulocytes or wheat germ, a CAT mRNA bearing an IRE in its 5'UTR (IRE-CAT) is translationally repressed in the presence of IRF. The wheat germ system which lacks endogenous IRF activity has been utilized to prove the repressor function of recombinant IRF expressed in *E.coli* (Gray et al., submitted). Addition of this factor specifically reduced the translation of IRE-CAT mRNAs in a dose-dependent manner.

The understanding of how iron regulates ferritin and eALAS translation posed the question whether binding of proteins to specific sites in the 5' UTR of an mRNA was generally sufficient to control mRNA translation. Since the IRE/IRF system currently represents the only physiological example for such a mechanism, surrogate experimental systems were developed using known RNA-binding proteins with physiological functions unrelated to eukaryotic translation (Stripecke and Hentze, 1992). The RNA-binding sites for the bacteriophage MS2 coat protein or the spliceosomal protein U1A were introduced into the 5'UTR of CAT indicator transcripts 31 nucleotides downstream from the cap structure. These mRNAs were efficiently translated in rabbit reticulocyte lysate and wheat germ extract in the absence of added cognate binding proteins. Addition of the purified binding proteins caused a dose-dependent, mRNA-specific repression of translation. Repression required high affinity binding ($K_d$ < 1 nM), as indicated by the analysis of mutated binding sites. This repression was reversible and did not alter the stability of the mRNA. When the same binding sites were placed 32 or 38 nucleotides further downstream, translational repression by RNA/protein complex formation was far less profound, indicating that the position-dependence originally described for the IRE/IRF system was a more general phenomenon.

Taken together, the results obtained with the IRE/IRF system and the surrogate *in vitro* systems using MS2 coat protein and U1A allow to formulate a hypothesis for translational regulation by mRNA/protein interactions (Melefors and Hentze, 1993): High affinity binding of a protein to an mRNA causes translational repression, provided that the binding site is located close (within ~40 nucleotides) to the cap structure. Mechanistically, repression appears to be sterical, because the effect of U1A and the MS2 coat protein suggests that no additional requirements other than high affinity mRNA-binding have to be met. It is currently not known whether (and under which circumstances) protein binding to sites within the open reading frame or the 3'UTR of an mRNA cause translational repression. Examples of *cis*-regulatory sequences which might serve as protein binding sites have been identified within the 5' and 3' UTRs of several translationally regulated mRNAs (Hyman and Wormington, 1988; Amaldi et al., 1989; Schäfer et al., 1990; Ahringer and Kimble, 1991; Chu et al., 1991; Kwon and Hecht, 1991; Kaspar et al., 1992), but their precise roles in mediating translational control require further definition.

If high affinity mRNA/protein complexes repress translation, how can this mechanism be utilized to achieve signal-dependent translational regulation? In the case of the IRE/IRF system, the regulatory signal (iron) controls the RNA-binding activity of the repressor protein post-translationally. More generally, biological signals may regulate the RNA-binding affinity of pre-existing *trans-*

acting repressor proteins. Alternatively, the availability of active repressors can be controlled at the level of their synthesis [by transcriptional or post-transcriptional mechanisms] or degradation. Finally, negative feedback autoregulation of the synthesis of RNA-binding proteins can be considered. The protein may partition between its "functional" binding site on a non-mRNA (e.g. small nuclear or small cytoplasmic RNA, ribosomal RNA, tRNA) and a binding site on its own mRNA. Under conditions of excess 'free' protein, it could bind to its mRNA and suppress further synthesis. Clearly, translational regulation by RNA/protein interactions represents a versatile biological control system capable of rapid responses to changing environmental signals or cellular requirements.

## References

AHRINGER, J. AND KIMBLE, J. (1991) Control of the sperm-oocyte switch in Caenorhabditis elegans hermaphrodites by the fem-3 3' untranslated region. Nature 349: 346-348.

AMALDI, F., BOZZONI, I., BECCARI, E. AND PIERANDREI-AMALDI, P. (1989) Expression of ribosomal protein genes and regulation of ribosome biosynthesis in Xenopus development. Trends Biochem. Sci. 14: 175-178.

AZIZ, N. AND MUNRO, H.N. (1987) Iron regulates ferritin mRNA translation through a segment of its 5' untranslated region. Proc. Natl. Acad. Sci. USA. 84: 8478-8482.

BEINERT, H. AND KENNEDY, M.C. (1989) Engineering of protein bound iron-sulfur clusters. A tool for the study of protein and cluster chemistry and mechanism of iron-sulfur enzymes. Eur. J. Biochem. 186: 5-15.

BROWN, P.H., DANIELS-McQUEEN, S., WALDEN, W.E., PATTINO, M.M., GAFFIELD, L., BIELSER, D. AND THACH R.E. (1989) Requirements for the translational repression of ferritin transcripts in wheat germ extracts by a 90-kDa protein from rabbit liver. J. Biol. Chem. 264: 13383-13386.

CASEY, J.L., HENTZE, M.W., KOELLER, D.M., CAUGHMAN, S.W., ROUAULT, T.A., KLAUSNER, R.D. AND HARFORD, J.B. (1988) Iron-responsive elements: regulatory RNA sequences that control mRNA levels and translation. Science 240: 924-928.

CAUGHMAN, S.W., HENTZE, M.W., ROUAULT, T.A., HARFORD, J.B. AND

KLAUSNER, R.D. (1988) The iron-responsive element is the single element responsible for iron-dependent translational regulation of ferritin biosynthesis. Evidence for function as the binding site for a translational repressor. J. Biol. Chem. 263: 19048-19052.

CHU, E., KOELLER, D.M., CASEY, J.L., DRAKE, J.C., CHABNER, B.A., ELWOOD, P.C., ZINN, S. AND ALLEGRA, C.J. (1991) Autoregulation of human thymidylate synthase messenger RNA translation by thymidylate synthase. Proc. Natl. Acad. Sci. USA. 88: 8977-8981.

CONSTABLE, A., QUICK, S., GRAY, N.K. AND HENTZE, M.W. (1992) Modulation of the RNA-binding activity of a regulatory protein by iron in vitro: Switching between enzymatic and genetic function? Proc. Natl. Acad. Sci. USA. 89: 4554-4558.

COX, T.C., BAWDEN, M.J., MARTIN, A. AND MAY, B.K. (1991) Human erythroid 5-aminolevulinate synthase: promoter analysis and identification of an iron-responsive element in the mRNA. EMBO J. 10: 1891-1902.

DANDEKAR, T., STRIPECKE, R., GRAY, N., GOOSSEN, B., CONSTABLE, A., JOHANSSON, H.E. AND HENTZE, M.W. (1991) Identification of a novel iron-responsive element in murine and human erythroid δ-aminolevulinic acid synthase mRNA. EMBO J. 10: 1903-1909.

GOOSSEN, B., CAUGHMAN, S.W., HARFORD, J.B., KLAUSNER, R.D. AND HENTZE, M.W. (1990) Translational repression by a complex between the iron-responsive element of ferritin mRNA and its specific cytoplasmic binding protein is position-dependent in vivo. EMBO J. 9: 4127-4133.

GOOSSEN, B. AND HENTZE, M.W. (1992) Position is the critical determinant for function of iron-responsive elements as translational regulators. Mol. Cell. Biol. 12: 1959-1966.

GRAY, N.K., CONSTABLE, A., QUICK, S. AND HENTZE, M.W. Translational control mediated by an iron-responsive element (IRE) and recombinant iron-regulatory factor (IRF) in vitro. Submitted.

HAILE, D.J., HENTZE, M.W., ROUAULT, T.A., HARFORD, J.B. AND KLAUSNER, R.D. (1989) Regulation of interaction of the iron-responsive element binding protein with iron-responsive RNA elements. Mol. Cell. Biol. 9: 5055-5061.

HAILE, D.J., ROUAULT, T.A., TANG, C.K., CHIN,J., HARFORD, J.B. AND KLAUSNER, R.D. (1992) Reciprocal control of RNA-binding and aconitase activity in the regulation of the iron-responsive element binding protein: Role of the iron-sulfur cluster. Proc. Natl. Acad. Sci. USA. 89: 7536-7540.

HENTZE, M.W., CAUGHMAN, S.W., ROUAULT, T.A., BARRIOCANAL, J.G., DANCIS, A., HARFORD, J.B. AND KLAUSNER, R.D. (1987) Identification of the iron-responsive element for the translational regulation of human ferritin mRNA. Science 238: 1570-1573.

HENTZE, M.W., CAUGHMAN, S.W., CASEY, J.L., KOELLER, D.M., ROUAULT, T.A., HARFORD, J.B. AND KLAUSNER, R.D. (1988) A model for the structure and function of iron-responsive elements. Gene 72: 201-208.

HENTZE, M.W., ROUAULT, T.A., HARFORD, J.B. AND KLAUSNER, R.D. (1989) Oxidation-reduction and the molecular mechanism of a regulatory RNA-protein interaction. Science 244: 357-359.

HENTZE, M.W. AND ARGOS, P. (1991) Homology between IRE-BP, a regulatory RNA-binding protein, aconitase, and isopropylmalate isomerase. Nucl. Acids Res. 19: 1739-1740.

HYMAN, L.E. AND WORMINGTON, W.M. (1988) Translational inactivation of ribosomal protein mRNAs during Xenopus oocyte maturation. Genes Dev. 2: 598-605.

KAPTAIN, S., DOWNEY, W.E., TANG, C., PHILPOTT, C., HAILE, D., ORLOFF, D.G., HARFORD, J.B., ROUAULT, T.A. AND KLAUSNER, R.D. (1991) A regulated RNA binding protein also possesses aconitase activity. Proc. Natl. Acad. Sci. USA. 88: 10109-10113.

KASPAR, R.L., KAKEGAWA, T., CRANSTON, H., MORRIS, D.R., WHITE, M.W. (1992) A regulatory cis element and a specific binding factor involved in the mitogenic control of murine ribosomal protein L32 translation. J. Biol. Chem. 267: 508-514.

KLAUSNER, R.D. AND HARFORD, J.B. (1989) Cis-trans models for post-transcriptional gene regulation. Science 246: 870-872.

KÜHN, L.C. AND HENTZE, M.W. (1992) Coordination of cellular iron metabolism by post-transcriptional gene regulation. J. Inorgan. Biochem. 47, 183-195.

KWON, Y.K. AND HECHT, N.B. (1991) Cytoplasmic protein binding to highly conserved sequences in the 3' untranslated region of mouse protamine 2 mRNA, a translationally regulated transcript of male germ cells. Proc. Natl. Acad. Sci. USA. 88: 3584-3588.

LEIBOLD, E.A. AND MUNRO, H.N. (1988) Cytoplasmic protein binds in vitro to a highly conserved sequence in the 5' untranslated regions of ferritin heavy and light subunit mRNAs. Proc. Natl. Acad. Sci. USA. 85: 2171-2175.

MELEFORS, Ö. AND HENTZE, M.W. (1993) Translational regulation by mRNA/protein interactions in eukaryotic cells: ferritin and beyond. Bioessays 15: in press.

MELEFORS, Ö., GOOSSEN, B., JOHANSSON, H.E., STRIPECKE, R., GRAY, N.K. AND HENTZE, M.W. (1993) Translational control of 5-aminolevulinate synthase mRNA by iron-responsive elements in erythroid cells. J. Biol. Chem. 268: in press.

MÜLLNER, E.W., NEUPERT, B. AND KÜHN, L.C. (1989) A specific mRNA binding factor regulates the iron-dependent stability of cytoplasmic transferrin receptor mRNA. Cell 58: 373-382.

PATINO, M.M. AND WALDEN, W.E. (1992) Cloning of a functional cDNA for the rabbit ferritin mRNA repressor protein. Demonstration of a tissue-specific pattern of expression. J. Biol. Chem. 267: 19011-19016.

QUICK, S., GOOSSEN, B., HIRLING, H., KÜHN, L.C. AND HENTZE, M.W. Expression of recombinant iron regulatory factor (IRF) in E. coli: A protein with two diverse, mutually exclusive biological activities. Submitted.

ROGERS, J. AND MUNRO, H. (1987) Translation of ferritin light and heavy chain subunit mRNAs is regulated by intracellular chelatable iron levels in rat hepatoma cells. Proc. Natl. Acad. Sci. USA. 84: 2277-2281.

ROUAULT, T.A., STOUT, C.D., KAPTAIN, S., HARFORD, J.B. AND KLAUSNER, R.D. (1991) Structural relationship between an iron-regulated RNA-binding protein (IRE-BP) and aconitase: functional implications. Cell 64: 881-883.

SCHÄFER, M., KUHN, R., BOSSE, F. AND SCHÄFER, U. (1990) A conserved element in the leader mediates post-meiotic translation as well as cytoplasmic polyadenylation of a Drosophila spermatocyte mRNA. EMBO J. 9: 4519-4525.

STRIPECKE, R. and HENTZE, M.W. (1992) Bacteriophage and spliceosomal proteins function as position-dependent cis/trans repressors of mRNA translation in vitro. Nucl. Acids Res. 20: 5555-5564.

THEIL, E.C. (1990) Regulation of ferritin and transferrin receptor mRNAs. J. Biol. Chem. 265: 4771-4774.

ZÄHRINGER, J., BALIGA, B.S. AND MUNRO, H.N. (1976) Novel mechanism for translational control in regulation of ferritin synthesis by iron. Proc. Natl. Acad. Sci. USA. 73: 857-861.

# FRAMESHIFTING IN PROKARYOTES AND EUKARYOTES

H. Engelberg-Kulka
Department of Molecular Biology
Hadassah Medical School
Hebrew University
Jerusalem
Israel

A.J. & S.M Kingsman
Department of Biochemistry
South Parks Road
Oxford
OX1 3QU
UK

Frameshifting has been observed in several systems to date. It is used to increase the genetic content of a stretch of DNA/RNA or to produce different proteins that have related regions towards their amino termini. The efficiency of these events also determines the dosage of the alternative gene products. In this article we will describe frameshifting systems from both prokaryotes and eukaryotes. The prokaryotic systems that will be described will emphasise those frameshifting events that are involved in cellular gene expression whereas the eukaryotic examples will be from retroviruses and retrotransposons where the molecular determinants of shifting have been studied in most detail( see Jacks,1989, for a full review).

## 1. TRANSLATIONAL FRAMESHIFTING IN RETROVIRUSES AND RETROTRANSPOSONS.

### Retroviruses

The gag and pol genes are adjacent at the beginning of all retroviral transcriptional units (Figure 1) and in many cases the 3' end of gag and the 5' end of pol overlap by up

NATO ASI Series, Vol. H 97
Post-transcriptional Control of Gene Expression
Edited by Orna Resnekov and Alexander von Gabain
© Springer-Verlag Berlin Heidelberg 1996

to a few hundred nucleotides with pol in the -1 reading frame with respect to gag. Both genes are translated from the full length 'genomic' RNA to produce two primary translation products, a GAG precursor protein and a GAG:POL fusion pr protein. The production of the fusion protein requires that the gag and pol reading frames are brought into translational phase: that is an adjustment of the ribosome's reading phase by a -1 shift is necessary. For many years it was assumed that the shift was accomplished by a small splice. However, in 1985 Jacks and Varmus (1985) showed that RSV frameshifting could be achieved in an in vitro translation system suggesting that the frameshift was due to some event at the ribosome.

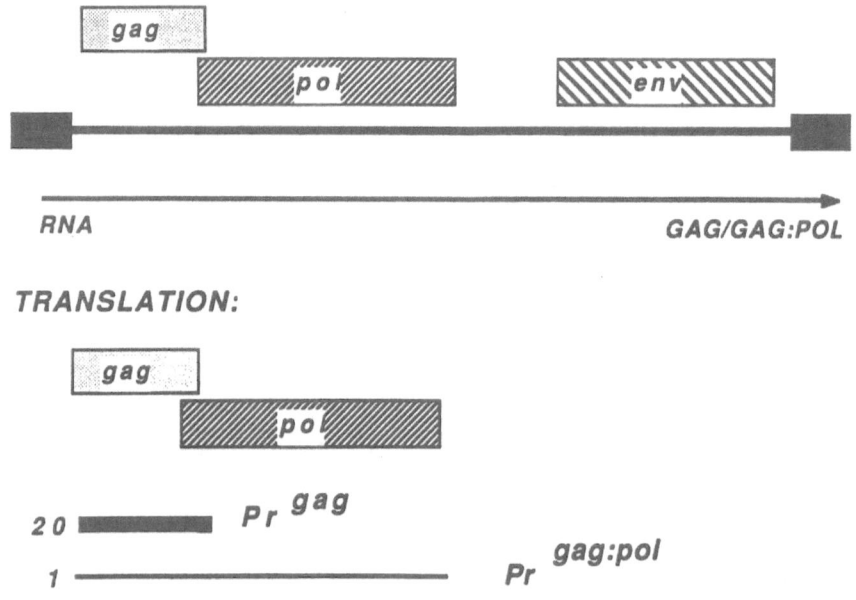

Figure 1. - The genetic organisation of a retroviral provirus and the expression of the gag and pol genes.

In HIV, for example gag and pol overlap by 241 nucleotides. (Ratner et al., 1985). The gag precursor protein, p55, is the primary product of simple translation of the full length viral RNA. The gag:pol fusion protein, p160, also a precursor, is the product of frameshifted translation of the same RNA. The efficiency of the shift is about 5% and therefore, in common with many other retroviruses, the relative abundance of the two precursors is 20:1, although this has only been determined accurately in vitro (Jacks et al., 1988a; Wilson et al., 1988).

The standard assay for frameshifting uses an in vitro translation reaction to translate an SP6 generated RNA (Jacks and Varmus, 1985). The RNA is constructed with a preshift sequence, a candidate shift site and then a post shift sequence. Shifting is detected either by the presence of a second higher molecular weight band in an immunoprecipitation with an antibody directed against the protein encoded by the preshift sequence or by expression of a repoter gene that is dependent on the frameshift. Most of the results that we describe for retroviruses were obtained using this assay although some studies have been carried out in vivo (Wilson et al., 1988; Parkin et al., 1992).

Frameshifting must occur in the gag/pol overlap region so that the shift into the -1 pol phase is achieved before the ribosome reaches the stop codon of the gag open reading frame. Jacks et al (1988a) showed that in HIV shifting occured at codon 3 of the pol open reading frame with the sequence TTT.TTA.GGG being read as PHE.LEU.GLY in GAG and as PHE.LEU.ARG in the GAG:POL fusion. However, the data reveal that there is a significant (at least 30%) amount of phenylalanine at the second position giving an alternative shift site sequence of PHE.PHE.ARG. These data were in agreement with the observation of Wilson et al (1988) who showed that the sequence required for HIV shifting was within the first 16 nucleotides of the overlap region.

The observation that such a short stretch of the overlap

region was required for shifting in HIV was surprising in the context of what was known about other retroviruses (Jacks et al., 1987; Jacks et al., 1988b). In almost every case where a virus makes use of frameshifting as a gene expression strategy the shift site is followed within a very short distance, usually less than 10 nucleotides, by a region of secondary structure. The secondary structure may be a simple stem loop or a pseudoknot (Jacks et al., 1988b; Brierley et al., 1989). A simple suggestion for the existence of these structures is that they slow the ribosome making it more likely that it will 'slip' at the shift site. In RSV, there is good evidence (Jacks et al., 1988b) that the downstream secondary structure is required from mutational studies that show that destabilisation of the stem reduces frameshifting efficiency and unpiublished data suggest that ribosomal pausing may occur at the stem-loop. Curiously, in HIV, although the stem-loop exists it does not lie within the first 16 nucleotiudes of the overlap and therefore it appears not to have a function in efficient shifting in vitro or in yeast (Wilson et al., 1988). It would seem therefore that HIV shifting does not require ribosomal stalling for the shift to occur.

Shift sites have been identified in several other retroviruses using the in vitro assay and, where the analyses are sufficiently complete, it seems that the functional sites are heptanucleotides. Examination of these sites reveals that they fall into two classes (Figure 2). In Class I shift sites the first 6 nucleotides are of the general sequence X.XXY.YY where the stops define the gag codons. Class II shift sites are of the general sequence X.XXX.XX. The difference between these two classes may explain differences in the requirement for stem-loop structures. In a class I shift site such as that for RSV there are two homopolymeric runs of 3 nucleotides. The phase relationship of these two triplets to the gag and pol reading frames is always the same irrespective of the sequence. In RSV the heptanucleotide sequence AAAUUUA is the shift site and is thought to mediate shifting through a -1 slip of the codon:anticodon interactions at both the A and P sites (Jacks et al., 1988b). Following the slip the tRNA$^{LEU}$ (UUA) and the tRNA$^{ASN}$(AAU) would be held on the RNA by 2 out of three base pairs each. Normal translocation would then take place and translation would proceed, in phase with pol. Such a mechanism can be brought about by any adjacent homopolymeric triplets as long as the distribution of the triplets with respect to the reading frames is the same as in Class I sequences. This mechanism does not allow a shift to the +1/-2 phase. In Class II, such as HIV, there is a single homoplymeric run of 6 nucleotides or two identical triplets. However, the relationship of these triplets to the gag and pol reading frames is the same as in Class I. In HIV the heptanucleotide sequence is UU.UUU.UA. Shifting to -1 can be achieved by a mechanism almost identical to that proposed for Class I. IN this case the tRNA$^{PHE}$(UUU) and tRNA$^{LEU}$(UUA) would both slip back one nucleotide in the A and P sites. The only difference between Class I and Class II would be that after the slip one of the tRNAs, tRNA$^{PHE}$, would be held on the RNA by 3 out of 3 base pairs. tRNA$^{LEU}$ would be held, as in Class I, by 2 out of 3 base pairs. In Class II, therefore, the slip is maintained by 5 base pairs as opposed to 4 in Class I. Consequently the slip may be more efficient. The amino acid sequence over the frameshift region of HIV would be predicted, by this model,is ASN-PHE-LEU-ARG, the sequence determined by Jacks et al. (1988a). However, slippage at the A and P sites is not the only mechanism open to a Class II shift sequence. It is possible that in HIV, for example, the tRNA$^{PHE}$(UUU) slips during translocation prior to the tRNA$^{LEU}$(UUA) entering the A site. In this case the tRNA$^{PHE}$ slips to -1, maintaining 3 out of 3 base pairs and exposing a free UUU codon in the A site. Rather than tRNA$^{LEU}$ entering the A site a second tRNA$^{PHE}$ enters and the -1 shift is completed. The slip is maintained by 6 out of 6 pairing and is likely, therefore to be quite stable. This mechanism predicts the amino acid sequence ASN-PHE-PHE-ARG accross the shift region. Both A and P site shifting and

translocation shifting seem equally plausible. In HIV, of
course, if both occured it would result in micro-heterogeneity
in the GAG:POL fusion protein. Close examination of the data
of Jacks et al (1988a) shows exactly that. As mentioned
previously at least 30% of the HIV shifts produce the
sequence ASN-PHE-PHE-ARG. This suggests that Class II sites
are much shiftier than Class I sites and this may be reflected
in differences in their respective requirements for downstream
secondary structures. The requirements for HIV shifting appear
to be simple. The virus makes use of a generally shifty
sequence to express its pol gene and this is all that is
required. In contrast, RSV, uses a somewhat less shifty
sequence and enhances its shiftyness by using a stem-loop
structure to pause the ribosome.

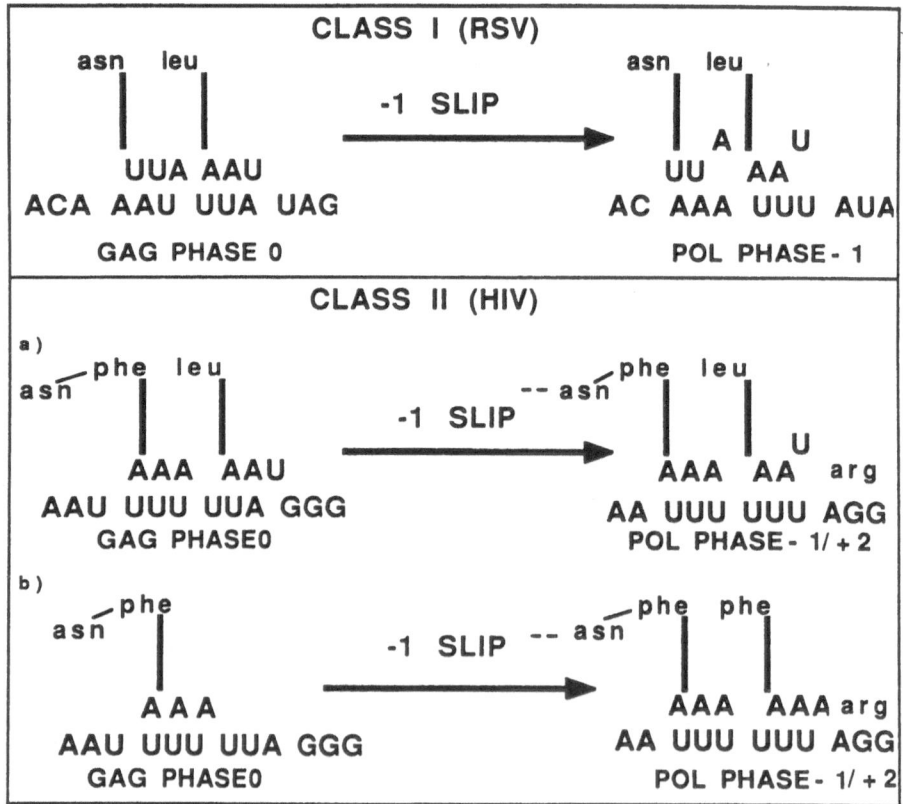

Figure 2. Frameshift mechanisms in Class I and Class II shift sites

The notion that there are two classes of retroviral shift site needs to be qualified. There are shift sites that do not conform to the class I/II pattern such as the MMTV pro:pol shift site. Also the Ty retrotransposon uses an entirely different shifting mechanism. In addition the observations made in vitro may not reflect the situation in vivo. Recent data from Parkin et al (1992) suggest that HIV shifting may be enhanced by the stem-loop structure in vivo although the data are not entirely conclusive.

**Retrotransposons**

The only retrotransposon in which frameshifting has been analysed in detail is the yeast Ty1 element. Ty1 is a movable DNA element that transposes via an RNA intermediate and a reverse transcriptase step. It is structurally and functionally very similar to a retroviral provirus (Kingsman and Kingsman, 1988). Ty1 has two open reading frame TYA and TYB that are gag and pol analogues respectively. TYA and TYB overlap by 40 nucleotides with TYB in the +1 phase with respect to TYA. Like retroviral gag and pol genes, TYA is expressed by simple translation and TYB is expressed as a TYA:TYB fusion protein via a ribosomal frameshift (Mellor et al., 1985). Remarkably, even though Ty1 is obviously related to retroviral proviruses this frameshift event is a +1 shift rather than -1. Mutation studies have located the frameshift determinant to the sequence CUU.AGG.C, where the codons represent the TYA frame. Amino acid sequence analysis accross this region in the frameshift product revealed that the shift occured at the CUUA sequence. This is mediated by tRNA$^{LEU}$(CUA) which can decode both CUU and UUA (Randerath et al., 1979). The model proposed by Belcourt and Farabaugh (1990) states that the tRNA$^{LEU}$ slips in the P site in contrast to the retroviral models which suggest A and P site slippage. This is based on the amino acid sequence of the shifted product which is THR.LEU.GLY rather than THR.LEU.ARG which would be the sequence if simultaneous slippage occured when the A site was occupied by tRNA$^{ARG}$.

Ty shifting, therefore, is quite different from retroviral shifting although, like some retroviral shift mechanisms an induced ribosomal stall is required for the P-site slippage to occur. In Ty this is mediated by the AGG ARG codon. The tRNA species for this codon is present at low concentrations in yeast and so as the ribosome encounters this codon in the A-site it slows down allowing time for the P-site slippage to occur. That this notion was correct was demonstrated elegantly by Becourt and Farabaugh (1990) who showed that if tRNA$^{ARG}$(AGG) is overexpressed in yeast then Ty frameshifting is suppressed. A curious feature of this system is that this induced stall is specific for tRNA$^{ARG}$(AGG) in that no other rare codon will do. This might suggest that the emchanism is more complex than simple stalling.

**Conclusions**

It is clear that both retroviruses and retransposons employ frameshifting as a strategy to compress genetic information and to control the relative levels of the gag/TYA and pol/TYB proteins. This strategy also ensures that the enzyme activities encoded by pol/TYB are packaged into the particles that are required for transposition or production of infectious virus because in precursor form they are fused to the self-assembling gag/TYA encoded proteins. In every case shifty sequences are used but, depending on the natural efficiecy of these sequences their shiftyness may be enhanced by other mechanisms. At present there is no evidence for any trans-acting proteins in these systems.

**2.  TRANSLATIONAL FRAMESHIFTING AND BYPASSING IN *ESCHERICHIA COLI*.**

Only three examples of cellular genes expressed by frameshifting have been characterized, all of them in *E. coli* (genes *prfB, dnaX,* and *trpR*).  We shall discuss all three here. We shall also compare the various mechanisms leading to frameshifting events in these three *E. coli* genes. It appears that the role of the frameshifting event differs in each case, and overcoming a termination codon is only one of several of its regulatory implications.

**Frameshifting of the *E. coli prfB* gene.**

The *E. coli prfB* gene codes for the protein release factor 2 (RF2) which directs translation termination in response to a UGA or a UAA termination codon. It has been shown that codon 26 is an in-frame UGA termination codon, and that in order to overcome it a +1 frameshift must occur permitting translation to continue into the +1 open reading frame. Translation finally terminates at a UAG termination codon when the complete 371 amino acid long RF2 molecule is synthesized (Craigen *et al.*, 1985; Craigen and Caskey, 1986, 1987).

As in other described programmed frameshifting events (see Section 1), two *cis* elements in the mRNA are involved in the case of RF2. The first is the slippery site itself (CUU UGA) where the +1 frameshifting event occurs. For most programmed frameshifting events the second cis element is a downstream stem and loop (or pseudoknot) structure (see Section 1). In contrast, in the case of RF2, this second element is a Shine-Dalgarno sequence (AGG-GGG) located several nucleotides upstream from the slippery site. Base pairing between this Shine-Dalgarno sequence and the 16S rRNA is required for RF2 frameshifting (Weiss *et al*, 1988).

The most interesting aspect of *prfB* frameshifting is its regulatory implications. It has been suggested that RF2 autoregulates its own production by the in-frame UGA termination codon found within the early coding region of *prfB* mRNA (Craigen *et al.*, 1985). When adequate RF2 is present, peptide chain termination would occur at this site, thus limiting its own production. On the other hand, when levels of RF2 are low, a readthrough/frameshift mechanism would permit increased expression of RF2.

## Frameshifting of the *E. coli dnaX* gene.

The *dnaX* gene of *E. coli* has only one open reading frame for a 71kD polypeptide from which two distinct DNA polymerase II holenzyme subunits, τ (71kD) and γ (47kD), are produced. The N-termini of the τ and γ subunits are identical. Originally, it was thought that the γ was generated by proteolytic cleavage. However, recently it has been shown that the generation of the γ subunit does not require proteolytic cleavage (Blinkowa and Walker, 1990; Flower and McHenry, 1990; Tsuchihashi and Kornberg, 1990). Rather, it is a result of a -1 translational frameshifting event of a remarkably high frequency (*i.e.* about 80% in an overproduced cell).

One important feature of *dnaX* frameshifting is that it provides no means for overcoming an in-frame termination codon. Instead, a UGA termination codon in the -1 frame causes the formation of the truncated 47kD polypeptide. Thus, there is an interesting distinction between frameshifting events in the *dnaX* on the one hand, and in *prfB* genes and retrovirus genes on the other (see Section 1). The generation of the γ subunit depends of termination, while the generation of reverse transcriptase and proteases in retroviruses, and RF2 in *E. coli*, depends on avoiding termination.

Another remarkable feature of _dnaX_ frameshifting is that the sites involved are structurally strikingly similar to the signals responsible for frameshifting in many retroviruses (see Section 1). As in many retroviruses, in *dnaX* the -1 slippery site is a stretch of six adenosine residues followed by a stimulatory site which is a stable RNA stem-loop structure. In constrast, however, the regions downstream from the stem-loop structure in dnaX are indispensable for frameshifting. Thus, *dnaX* frameshifting does not involve the structural element of the "pseudoknot."

**Frameshifting in the expression of the *E. coli trpR* gene.**

The *E. coli trpR* gene codes for the *trp* repressor which regulates the transcription initiation of four *E. coli* operons and genes involved in tryptophan metabolism and transport for review see Somerville, 1992). The 108 amino acid long 12kD trp repressor is encoded by the 324 nucleotide long open reading frame of the <u>trpR</u> gene. A +1 frameshifting event occurs during the expression of *trpR* (Benhar *et al.*, 1990; 1992). As a result, an additional 10kD +1 product is synthesized which is shorter than the 12kD frame 0 product. We have shown that the *trpR* +1 frameshift product has the same amino acid composition at its N-terminus as does the *trpR* frame 0 product, but differs from it in its C-terminus.

One of the experimental systems that we used for characterizing *trpR* frameshifting was a *trpR-lac`Z* fusion system (see Figure 3A); we fused the segment of the *trpR* downstream from codon 78, in either its 0 or +1 frame to the 10th codon of the *lac`Z* gene. We called the resulting gene fusions *trpR$_0$-lac`Z*. Using this experimental system, we found that the level of expression of the trpR +1 reading frame is about 5% of its expression in the 0 frame (Benhar *et al.*, 1990; 1992). We used this experimental system to define the trpR mRNA sequences required for trpR frameshifting. In this study, we identified and mapped the shortest sequences required for *trpR$_{+1}$-lac`Z* frameshifting by deletion and replacement experiments (Benhar and Engelberg-Kulka, 1992). The results of these experiments indicated that the frameshifting process requires the trpR nucleotide sequence from codon 54 to codon 70. However, the region between codons 54 to 65 of trpR could be replaced by an unrelated sequence which must be longer than 10 translatable codons long if frameshifting is to occur. Thus, *trpR$_{+1}$-lac`Z* frameshifting requires two adjacent *cis* elements: i) A specific sequence of *trpR* between codons 65 and 70 (or perhaps just part of it) which must be preceded by: ii) A non-specific open reading frame longer than 10 translatable codons (Figure 3B). It has

been reported that ribosomes protect about 35 nucleotide of
mRNA from nucleolytic digestion (Kang and Cantor, 1985).
Thus, an initiating ribosome protects about 18 nucleotides (6
codons).    Since $trpR_{+1}$-$lac$`$Z$ frameshifting requires the
translation of more than 10 codons, we conclude that
frameshifting is permitted only during the elongation stage of
protein synthesis.    Moreover, using the same $trpR_{+1}$-$lac$`$Z$
fusion system, we have recently discovered that $trpR$ +1
frameshifting does not occur through a slippage of a single
nucleotide, as has been described for most examples of
frameshifting (Varmus, 1988; Parker, 1989).    Instead, the
transition from the 0 to the +1 frame of $trpR$ occurs by the
bypassing of a long segment of the $trpR_{+1}$-$lac$`$Z$ mRNA (Benhar
and Engelberg-Kulka, 1992).    This was revealed by the amino
acid sequence of the internal region of the $trpR_{+1}$-$lac$`$Z$ fusion
protein product.    For these experiments, we used a procedure
which we had developed for amino acid sequencing in internal
regions of proteins.    This procedure is based on cloning a
short oligodeoxynucleotide coding for the cleavage site of a
specific protease (factor Xa) in the middle of a gene.
Digestion of the isolated protein product of that gene with
the protease facilitates the direct determination of the amino
acid sequence of the region of the protein immediately after
the inserted proteolytic site.    Using this procedure, we had
already shown that the amino acid sequence of the $trpR_0$-$lac$`$Z$
fusion protein product matched the expected sequence deduced
from the coding nucleotide sequence (Benhar and Engelberg-
Kulka, 1991).    However, in the case of the $trpR_{+1}$-$lac$`$Z$ fusion,
the determined amino acid sequence was not co-linear with the
coding nucleotide sequence (Figure 3C).    The results of these
experiments showed that the $trpR_{+1}$-$lac$`$Z$ fusion is expressed
by the bypassing of a 55 nucleotide long gap.    This gap
includes the region between the $trpR$ codon 65, ATG, and a TTA
codon which is the third $lac$`$Z$ codon following the fusion
point between the genes. We also characterized the transition
from frame 0 to frame +1 occurring during the expression of

the *trpR*₊₁-*lac`Z* fusion by mutagenesis experiments. Replacing a sense codon by a stop codon in each different position of the +1 frame of the gap had no affect on the expression of *trpR*₊₁-*lac`Z*. In addition, we could not detect any spliced *trpR*₊₁-*lac`Z* mRNA molecules (Benhar *et al.*, 1992). Thus, *trpR*₊₁-*lac`Z* bypassing is not at the level of RNA splicing and we conclude that it is a translational event. Until now, the only other single well characterized example of a large gap being bypassed by the translational apparatus has been the bypassing of a 50 nucleotide gap (resulting in a -1 frameshifting event) involved in the expression of bacteriophage T4 gene *60* which encodes the 18kD subunit of DNA topoisomerase (Huang *et al.*, 1988; Weiss *et al.*, 1990). Thus, here we can discuss *trpR*₊₁- *lac`Z* bypassing only in relation to T4 gene *60* bypassing. In both systems, bypassing is not at the level of RNA processing but seems rather to be a translational event. On the other hand, in several other respects the two systems are completely different: i) In T4 gene *60* bypassing, the translating ribosomes move from frame 0 to frame -1 with an efficiency of nearly 100%: Thus, there is no translation in frame 0 through the gap or downstream from it (Weiss *et al.*, 1990). In contrast, in *trpR*₊₁-*lac`Z* only about 5% of the translating ribosomes bypass the gap into the +1 frame, while 95% of the translating ribosomes continue in frame 0; ii) For efficient bypassing to occur in T4 gene *60*, 16 <u>specific</u> amino acids are required in the nascent peptide which must be synthesized the by translation of the region before the gap; for bypassing to occur in *trpR*₊₁-*lac`Z*, translation of a <u>non-specific</u> segment, longer than 10 codons, preceding the gap is required. It has been suggested that in the case of T4 gene *60* the nascent peptide might interact with the ribosomes thereby enhancing the efficiency of bypassing (Weiss *et at.*, 1990). The much lower efficiency of bypassing the in *trpR*-*lac`Z* system could be explained by the lack of such an interaction by some auxiliary peptide; iii) The minimal essential site identified in *trpR*₊₁-*lac`Z* is not

similar to the sequences involved in T4 gene *60* bypassing;
iv) For T4 gene *60* bypassing, a stop codon in frame 0 must be
included in the stem-loop structure at the 5' junction of the
gap. There is no parallel element required at the 5' junction
of the gap in *trpR-lac`Z*; v) In T4 gene *60*, matched GGA codons
are present both before and after the gap (Weiss, 1990). The
codon at the 5' end of the gap has been called the take-off
site and the matched codon on the 3' end of the gap the
landing site. This matched pair could be replaced by a
different pair of matched codons without affecting the process
of bypassing the gap, suggesting that a single tRNA reads the
codons of either sides of the gap. In *trpR$_{+1}$-lac`Z,* the
"landing" site is a TTA leucine codon; there is no TTA codon
following the ATG methionine codon preceding the gap (Figure
1C). Thus, unlike the gap in T4 gene *60* (Weiss *et al.*, 1990),
the *trpR$_{+1}$-lac`Z* gap is not bordered by matched codons, and
*trp$_{+1}$-lac`Z* bypassing does not seem to be mediated by a single
tRNA which recognizes both the "take-off" and the "landing"
sites (Benhar and Engelberg-Kulka, 1992).

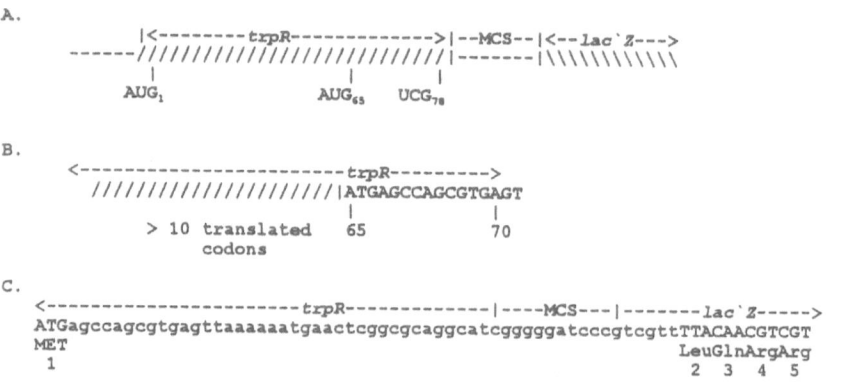

**Figure 3**    *trpR-lac`Z bypassing*.   (A) Schematic representation of the
*trpR-lac`Z* fusion. This fusion contains the *trpR* open reading frame, from
codon 1 to 78 of the gene (as indicated in the diagram), followed by a
multiple cloning site (MCS) of 9 nucleotides fused to the 10th codon of
*lac`Z*.   Codon 65 of *trpR* (coding for methionine) is also indicated as a
reference point.   (B)   Schematic representation of the sequences required
for *trpR$_{+1}$-lac`Z* frameshifting.   These sequences consist of a non-specific
5' end longer than 10 codons followed by the region between codons 65 to
70 of the *trpR* gene.   (C) Alignment of the *trpR$_{+1}$-lac`Z* amino acid sequence
with the coding nucleotide sequence.

We have suggested that the bypassing mechanisms of both T4 gene *60* and of *trpR*$_{+1}$-*lac`Z* both operate during translation but differently. If *trpR*$_{+1}$-*lac`Z* bypassing is to occur, the untranslated region of the mRNA must by looped out, thus bringing the "take-off" and "landing" sites into close proximity (Benhar and Engelberg-Kulka, 1992). Ribosome modeling studies have indicated that a ribosome can accommodate a folded mRNA loop of a considerable size (possibly more than 55 nucleotides) during translation without unfolding it in the process (Ada Yonat, personal communication). Theoretically, there may be three possible causes for such a looping out event; i) Internal base pairing within the mRNA sequence of the gap; ii) Base pairing of the mRNA gap sequence with 16S rRNA; or iii) The specific folding of the mRNA gap by a specific protein(s). We have excluded the first two possibilities by computer analysis of the gap sequence (including the required *trpR* sequences, codons 65-70). On the other hand, we find the possible interaction of a specific protein(s) in *trpR*$_{+1}$-*lac`Z* bypassing an attractive model. Such a protein(s) could even be one of the regular participants in translation like a ribosomal protein(s). Protein(s), otherwise engaged in the process of protein synthesis could occasionally interact with the mRNA sequences involved in *trpR*$_{+1}$-*lac`Z* bypassing. This would result in an occasional looping out of the gap causing it to be bypassed. This model could explain how a low level of *trpR*$_{+1}$-*lac`Z* bypassing (about 5%) would not interfere with the regular translation of frame 0 of the mRNA. In that respect, *trpR*$_{+1}$-*lac`Z* bypassing differs from other reported examples of frameshifting or bypassing, in which expression beyond the shift point is exclusively in the shifted frame (Parker, 1989; Weiss *et al.*, 1990).

Translational bypassing is a newly discovered mechanism of gene expression, and *trpR* is the first known cellular gene for which this mechanism seems to operate. This mechanism may be more generally used. The example of *trpR* and the high efficiency bypassing in T4 gene *60* may be the first clues to the existence of "translational introns" in other genes as well.

## REFERENCES

Atkins JF, Weiss RB and Gesteland RF (1990) Ribosomal gymnastics - degree of difficulty 9.5, style 10. Cell 62:413-423

Belcourt, M.F. and Farabaugh, P.J. (1990) Ribosomal frameshifting in the yeast retrotransposon Ty: tRNAs induce slippage on a 7 nucleotide minimal site. Cell 62, 339-352.

Benhar I, Miller C and Engelberg-Kulka H (1990) Frameshift-ing in the expression of the trpR gene of Escherichia coli. In: Post-transcriptional Control of Gene Expression. McCarthy JEG and Tuite MF (ed.s) Springer-Verlag Berlin pp591-602

Benhar I and Engelberg-Kulka H (1991) A procedure for amino acid sequencing in internal regions of proteins. Gene 103:79-82

Benhar I, Miller C and Engelberg-Kulka H (1992) Frameshift-ing in the expression of the Escherichia coli trpR gene. Mol Microbiol 6:2777-2784

Benhar I and Engelberg-Kulka H (1992) Frameshifting in the expression of the Escherichia coli trpR gene occurs by the bypassing of a segment of its coding sequence. Cell (in press)

Blinkowa A and Walker JR (1990) Programmed ribosomal frame-shifting generates the E. coli RNA polymerase III subunit from within the τ subunit reading frame. Nuc Acid Res 18:1725-1729

Craigen WJ, Cook RG, Tate WP and Caskey T (1985) Bacterial peptide chain release factors: Conserved primary structure and possible frameshift regulation of release factor 2. PNAS 82:3616-3620

Craigen WJ and Caskey CT (1986) Expresison of peptide chain release factor 2 requires high-efficiency frameshift. Nature 322:273-275

Craigen WJ and Caskey CT (1987) The function, structure and regulation of E. coli peptide chain release factor. Biochimie 69:1031-1041

Flower AM and McHenry CS (1990) the γ subunit of DNA poly-merase III holoenzyme of E. coli is produced by ribosomal frameshifting. PNAS 87:3713-3717

Huang WM, Ao SZ, Casjens S, Orlandi R, Zeikus R, Weiss R, Winge D and Fang M (1988) A persistent untranslated sequence within bacteriophage T4 DNA topoisomerase gene 60. Science 239:1005-1013

Jacks, T. (1989) Translational suppression in gene expression. Curr. Topics in Microbiol. Immunol. 157, 93-124.

Jacks, T. and Varmus, H. (1985) Expression of the Rous sarcoma virus pol gene by ribosomal framehifting. Science 230, 1237-1242.

Jacks, T., Power,M.D., Masiarz, F.R., Luciw, P.A., Barr, P.J. and Varmus, H. (1988a) Characterisation of ribosomal frameshifting HIV gag-pol expression. Nature 331, 280-283.

Jacks, T., Madhani, H.D., Masiarz, F.R., Varmus, H. (1988b) Signals for ribosomal frameshifting in the Rous sarcoma virus gag-pol region. Cell 55, 447-458.

Kang C and Cantor CR (1985) Structure of ribosome-bound messengerRNA as revealed by enzymatic accessibility. J Mol Biol 181:241-251

Kingsman, A.J. and Kingsman, S.M. (1988) Ty, a retroelement moving forward. Cell 53, 333-335.

Mellor , J., Fulton, S., Dobson, M.J., Wilson, W., Kingsman, S.M. and Kingsman, A.J. (1985) A retrovirus-like strategy for expression of a fusion protein encoded by the yeast transposon Ty1. Nature 313, 243-246.

Parker J (1989) Errors and alterations in reading the universal genetic code. Microbiol Rev 53:273-298

Parkin, N.T., Chammorro, M. and Varmus, H.E. (1992) Human immunodefieciency viris type 1 gag-pol frameshifting is dependent on downstream mRNA secondary structure: demonstration by expression in vivo. J.Virol. 66, 5147-5151.

Randerath, E., Gupta, R.C., Chia, L.L.S.Y., Chang, S.H. and Randerath, K. (1979) Yeast tRNA. Purification, properties and determination of nucleotide sequence by radioactive derivative methods. Eur. J. Biochem. 93, 79-94.

Ratner, L., Haseltine, W., Patarca, R., Livak, K.J., Starcich,B., Josephs, S.F., Doran,E.R., Rafalski, J.A., Ghrayeb, J., Chang, N.T., Gallo, R.C. and Wong-Staal, F. (1985) Complete nucleotide sequence of the AIDS virus HTLV-III. Nature 313, 277-284.

Somerville R (1992) The Trp repressor. A ligand-activated regulatory protein. Progress in nucleic acid research and molecular biology 42:274

TsTsuchihashi A and Kornberg A (1990) Translational frameshifting generates the $\gamma$ subunit of DNA polymerase II holoenzyme. PNAS 87:2516-2520

Weiss RB, Dunn DM, Dahlberg AE, Atkins JF and Gesteland RF (1988) Reading frame switch caused by base-pair formation between the 3' end of 16S rRNA and the mRNA during elongation of protein synthesis in *Escherichia coli* EMBO J.7:1503-1507

Weiss RB, Huang WM and Dunn DM (1990) A nascent peptide is required for ribosomal bypass of the coding gap in bacteriophage T4 gene *60*. Cell 62:117-126

Wilson,W., Braddock, M., Adams, S.E., Rathjen, P.D., Kingsman, S.M. and Kingsman, A.J. (1988) HIV expression strategies: ribosomal frameshifting is directed by a shortsequence in both mammalian and yeast systems. Cell 55, 1159-1169.

# THE CONTROL, DETECTION AND EXPLOITATION OF TRANSLATIONAL ERRORS

Mick F Tuite
Biological Laboratory,
University of Kent,
Canterbury,
Kent,
CT2 7NJ,
United Kingdom.

## WHAT ERRORS AND HOW FREQUENT?

Translational decoding of an mRNA does not occur without some errors in the amino acid sequence of the encoded polypeptide. While such translational errors have little genetic consequence, they can nevertheless be problematical to the cell. The translational machinery of both prokaryotes and eukaryotes has therefore evolved to keep errors to a minimum. In some cases, particularly in prokaryotic and viral systems, organisms have not only learnt to live with such errors, but to actually exploit them to either expand the decoding potential of their genomes or to regulate gene expression at the translational level. In this brief review I will describe the major types of decoding error that occur, how they can be detected and what causes them. In addition I will illustrate how some organisms have learnt to exploit these errors.

Missense errors result from the misincorporation of an incorrect amino acid into a growing polypeptide and may arise either from the misacylation of a tRNA or by incorrect codon-anticodon alignment during ribosomal decoding. Missense errors have an unpredictable consequence on the cell since single amino acid alterations can be silent or can completely inactivate the encoded gene product if they occur, for example, in the active site of an essential enzyme. That such errors do occur naturally is without question, but studies in E. coli suggest that missense error rates are of the order of $10^{-4}$ to $10^{-3}$ per codon translated the exact error rate for a given codon being dependent on the

NATO ASI Series, Vol. H 97
Post-transcriptional Control of Gene Expression
Edited by Orna Resnekov and Alexander von Gabain
© Springer-Verlag Berlin Heidelberg 1996

nucleotide context in which it is found, the abundance of the mRNA and whether or not it is a rare codon (Kirkwood et al., 1986). In eukaryotic cells estimates for natural missense error rates range from $10^{-5}$ to $10^{-4}$ per codon translated (Pollard, 1984) and while this would suggest a lower inherent translational error rate for eukaryotic cells, it is based on far fewer independent estimates than are available for E. coli.

Nonsense errors are of two types - ribosome "drop off" in which the ribosome prematurely terminates prior to the natural reading frame termination codon (UAA, UAG or UGA); or termination codon "by-pass" in which the natural reading frame termination codon is translated as if it were a sense codon, the ribosome continuing to translate into the 3' untranslated region (3'UTR) of the mRNA until it reaches the next in-frame termination codon. This latter type of error may occur as a consequence of defective termination release factor (RF) function or due to the presence of a tRNA which can decode one or other of the termination codons either through cognate or non-cognate codon-anticodon interaction; such tRNAs are usually referred to as nonsense suppressor tRNAs (Murgola, 1985). As will be described below naturally-occurring nonsense suppressor tRNAs have been described in a range of prokaryotes and eukaryotes.

Frameshift errors result from ribosomes entering a new reading frame during translation elongation and are generally +1 or -1 changes with respect to the natural reading-frame. This type of error can occur at random during ribosome translocation although there are now an increasing number of examples of "programmed" frameshifting events in which ribosomes shift frame during the translation of a specific mRNA at a fixed position and rate (Atkins et al., 1990).

Translational errors can therefore be viewed primarily as an irritation to the cell which have to be controlled, yet in some specific instances they may actually be beneficial to the cell. Translational errors can therefore be classified as either programmed or non-programmed biological events.

## DETECTING AND QUANTIFYING MISSENSE AND NONSENSE ERRORS

In growing cells translational errors are comparatively rare events and thus difficult to quantitate. Errors can be detected if one artificially elevates their frequency to a measurable level or assays must be devised which are exquisitely sensitive to one or other of the specific types of error. Table 1 indicates several ways in which the fidelity of translation in living cells can be experimentally reduced. Most widely utilised are error-inducing antibiotics and fidelity mutants.

---

- Error-inducing antibiotics
  - e.g. streptomycin, neomycin - E. coli
  - e.g. hygromycin B, paromomycin - yeast, animal cells

- Fidelity mutants
  - e.g. rpsD (ramA), tufB, relA - E. coli
  - e.g. sup35, sup45 - yeast

- Rare codons (low codon bias)

- Amino acid starvation

- Altering balance of translation factors
  - e.g. EF-Tu, RF - E. coli
  - e.g. EF-1α - yeast

---

**Table 1:** **Experimental means of elevating translational error rates in cells above the endogenous level.**

Error-inducing antibiotics have their effect via an interaction with the small ribosomal subunit e.g. streptomycin with the 30S subunit in E. coli, paromomycin with the yeast 40S subunit. Such interactions appear to perturb the fidelity of the codon-anticodon interaction and the widely used antibiotics (Table 1) are effective both in vivo and in vitro causing a wide spectrum of errors (Gorini, 1974). Transitional fidelity mutants of both E. coli and yeast (Saccharomyces cerevisiae) are hypersensitive to such error-inducing antibiotics presumably because they are unable to control the potentially lethal high levels of antibiotic-induced translational error. By-and-large antibiotics that induce translational errors in prokaryotes fail to do so in eukaryotic cells, and _vice versa_ reflecting the structural and functional differences between the ribosomes of prokaryotes and eukaryotes.

Assaying Nonsense Suppression.  Translation of a termination codon is almost invariably mediated by a specific nonsense suppressor tRNA (Murgola, 1985).  Such tRNAs can be assayed for in _in vitro_ translation systems by programming the system with an mRNA template with a known terminator and then assaying for the synthesis of a C-terminally extended polypeptide in the presence of the suppressor tRNA.  This assay has been widely used to detect naturally-occurring nonsense suppressor tRNAs (Hatfield _et al._, 1990).  Nonsense suppression can be detected _in vivo_ as the ability to simultaneously suppress the mutant phenotype of two or more defined nonsense mutations, a strategy extensively exploited in defining nonsense suppressor tRNAs in both _E. coli_ (Murgola, 1985) and yeast (Hinnebusch and Liebman, 1991).  An _in vivo_ assay requires that the amino acid inserted by the suppressor tRNA in response to the termination codon must be compatible with the function of the encoded polypeptide(s) whose activity is to be assayed.  This problem can be overcome by exploiting plasmid-borne gene fusions in which an upstream promoter and associated coding sequence is fused in-frame to a suitable reporter gene (e.g. _lacZ_, _cat_, _gus_) _via_ an in-frame termination codon.  We have successfully used this approach in _S. cerevisiae_ using a _PGK1_-_lacZ_ gene fusion to quantitate both nonsense suppressor tRNA efficiency and to quantify the endogenous termination readthrough error (Firoozan _et al._, 1991).  In such assay systems the nature of the amino acid inserted by the suppressor tRNA should have little or no effect on the activity of the reporter gene product.

Assaying Missense Suppression.  This type of error is much more difficult to detect and quantify.  _In vitro_ translation systems programmed with poly(U) have been used to assay for the mistranslation of the UUU (Phe) codon as a UU_A_, UU_G_ or _C_UU (Leu) codon and this has proven an effective means of identifying error-inducing antibiotics or for demonstrating that fidelity mutants contain error-prone ribosomes. These _in vitro_ systems are rarely physiologically relevant since they have a high endogenous Leu for Phe missense error rate due in part to the requirement for supraoptimal $Mg^{2+}$ concentrations necessary to force translation to initiate on the AUG-less poly(U) template.

Effective _in vivo_ assays for missense detection are limited, but at least four different approaches have been used principally in studies with _E. coli_:

(i) defining the thermolability of an easily assayable reporter enzyme e.g. ß-galactosidase; an increase in the rate of missense errors correlates with an increase in the thermolability of the ß-galactosidase (Hughes, 1991).

(ii) restoration of activity to a mutationally inactive reporter enzyme by misincorporation of an amino acid which restores partial or complete activity. This approach has been used with ß-lactamase, tryptophan synthetase and luciferase (Cornut and Willson, 1991).

(iii) misincorporation of $[^{35}S]$-Met or $[^{35}S]$-Cys into a polypeptide that can be easily purified, but does not usually encode Meto or Cys e.g. flagellin of _E. coli_ (Edelmann and Gallant, 1977).

(iv) alterations in the isoelectric point (pI) of a protein as detected by 2D SDS-PAGE. This so-called "stuttering" assay has been successfully used in both bacterial and mammalian cells and is very sensitive (Pollard, 1984). For example, misincorporation of a single glutamine (CAA, CAG) at a histidine (CAU, CAC) codon results in a net reduction in a protein's pI.

The most direct method of defining the extent and type of missense error for a given protein is to directly sequence the protein. However, this is a high cost, low throughput and low sensitivity option which has rarely been effective as a quantitative tool. The only exception has been in situations in which high level mistranslation is observed e.g. during the translation of a heterologous mRNA in _E. coli_ which contains a high proportion of "rare" codons (Scorer _et al_., 1991).

## MAINTAINING TRANSLATIONAL ACCURACY

Protein synthesis is a complex process which involves a large number of finely controlled translation factors interacting with the translating ribosome to direct synthesis of the encoded polypeptide chain. The ribosome plays a central role in ensuring the accuracy of the decoding process, a role uncovered initially by genetic studies in E. coli. The isolation of mutants in which the accuracy of translation was altered pinpointed three proteins of the (30S) ribosomal subunit, namely S4, S5 and S12 (Piepersberg et al., 1980). These three ribosomal proteins are thought to interact to define the degree of binding of an aa-tRNA to the A site of the ribosome (Kurland and Ehrenberg, 1985) since mutations within these proteins can alter the strength of binding of the aa-tRNA to the ribosome which in turn can cause loss of accuracy during decoding at the ribosomal A site.

Confirmation of an equivalent role for the eukaryotic homologues of these three ribosomal proteins in controlling accuracy of the yeast ribosome has come from a study of mutants which allow ribosomes to translate termination codons at high efficiency. Three of these so-called omnipotent suppressors arise from mutations in the yeast S4, S5 and S12 ribosomal protein equivalents, designated S13, S4 and S28 respectively (Hinnebusch and Liebman, 1991). Mutations in these genes have a number of phenotypes indicative of loss of ability to control induced translational errors. In vitro translation studies have confirmed that the ribosomes of these mutants are error-prone.

The ribosome does not act in isolation in ensuring optimal accuracy during translation. Studies, first in E. coli and subsequently in yeast, have also uncovered a role for the elongation factor responsible for bringing tRNA to the ribosome: EF-Tu in E. coli (Hughes, 1991), EF-1α in yeast (Sandbanken and Culbertson, 1988). Mutations which alter the structure of these functionally equivalent translation factors can result in a loss of accuracy. Altering the cellular concentrations of the wild-type factors can also profoundly influence the endogenous translational error rate. There are several other examples where alteration of the cellular levels of a specific translation factor affects the accuracy of translation. These data suggest that the process of ensuring optimal translational fidelity requires the finally balanced contribution of a number of components of the translational apparatus in addition to the ribosome.

Studies with omnipotent suppressors in yeast have also lead to the identification of two ribosome-associated translation factors designated Sup35p and Sup45p (Hinnebusch and Liebman, 1991). These factors appear to have no functional homologues in E. coli and while mammalian Sup35p and Sup45p homologues have recently been discovered there is as yet no evidence of a role in controlling translational accuracy in higher eukaryotic cells. One of the factors, Sup35p, shows significant sequence homology to EF-1α suggesting that it may interact with tRNA.

The fidelity of translation also relies on accurate aminoacylation of tRNAs prior to their being brought to the ribosome by EF-1α (EF-Tu). Such accuracy relies on the ability of aa-tRNA synthethases to discriminate between structurally similar tRNA molecules in competition with other aa-tRNA synthethases. This discrimination is thought to be achieved by interactions between the aa-tRNA synthetase and crucial nucleotides or "identity determinants" within the tRNA molecule (Normanly et al., 1986).

## EXPLOITING TRANSLATIONAL ERRORS

Not all translational errors are simply mistakes which are triggered by the ribosome or one of its associated translation factors. There is now increasing evidence for the exploitation of certain non-standard mRNA decoding events in viruses, bacteria, yeast, plants and mammals (Atkins et al., 1990; Parker, 1989). Such exploitation is restricted to a few specific mRNAs, but affords a novel means of translational regulation causing ribosomes to switch reading frame (i.e. frameshifting) or to by-pass termination codons (i.e. natural suppression) or to simply by-pass whole regions of an mRNA (i.e. hopping). With

| Translation Factor | Function |
|---|---|
| Ribosomal proteins | tRNA binding<br>translocation<br>codon:anticodon alignment |
| Elongation factors | |
| •     EF-1α (EF-Tu)<br>•     EF-2 (EF-G)<br>•     EF-3 | binds aa-tRNA to ribosome<br>translocation; reading frame<br>maintenance<br>codon:anticodon alignment(?) |
| Ribosome-associated factors<br>e.g. Sup35p, Sup45p | ancillary translation factors<br>codon:anticodon alignment(?) |
| Aminoacyl-tRNA Synthetases | charging tRNAs |

**Table 2: Protein synthesis factors important for maintaining translational fidelity**

the exception of hopping, these programmed errors also form part of the repertoire of non-programmed errors described above.

Frameshifting. Two basic types of programmed frameshifts have been described, namely the +1 and -1 reading frameshifts (Atkins et al., 1990; 1991). These events occur at a specific sequence within an mRNA at a site which contains a run four or more identical bases and is usually stimulated by one or more mRNA structures e.g. pseudoknots, located 3' to "shifty sequence". The best characterised examples of programmed +1 frameshifts are those that occur within the mRNA encoding E. coli polypeptide chain release factor RF2, and the +1 frameshift in the yeast transposon Ty1 which occurs at the boundary of the TyA-TyB coding sequence and is required to ensure in-frame translation of the TyB coding sequence by ribosomes initiating at the AUG of the TyA sequence. The -1 frameshift has been widely studied in retroviruses (e.g. HIV, MMTV) and coronaviruses (e.g. infectious bronchitis virus - IBV) and appears to arise via a common mechanism involving a restricted number of "slippery sites" which usually contain $A_6$, $U_6$ or $G_3A_3$ runs at the site of the frameshift.

In many cases the frequency of programmed frameshifting can

approach 50%. A preponderance of examples of frameshifting in viral or prokaryotic mRNA indicate that this mechanism for regulating gene expression at the level of translation is favoured where genomes are small and genetically compact and there is a need to optimise the use of the limited genetic storage capability.

<u>Natural Suppression and Termination Codon Readthrough</u>. The rapid acquisition of gene sequence data has revealed that in the nuclear genes of some organisms e.g. certain Protozoa, and in the mitochondrial genomes of the majority of organisms studied, one or more termination codons have become sense codons (Fox, 1985). There is also the unusual case of the UGA codon being decoded as selenocysteine, "the 21st amino acid", in certain mRNAs e.g. in <u>E. coli</u> formate dehydrogenase and certain mammalian glutathione peroxidases (Böck <u>et al</u>., 1991). As with programmed frameshifting, sequences 3' to the "special" UGA codon appear to be required for this unusual translational event, while studies in <u>E. coli</u> have also indicated that one or more specialised translation factors are required.

As was first noted over 20 years ago all three termination codons are not 100% efficient at signalling termination during translation i.e. they are leaky. This finding has led to the discovery of a number of naturally-occurring tRNAs in bacterial, plant and mammalian cells which are able to decode one or other of the 3 termination codons, all be it inefficiently (Atkins <u>et al</u>., 1990; Hatfield <u>et al</u>., 1990). Such tRNAs insert either glutamine, leucine, serine, tyrosine or tryptophan depending on the identity of the tRNA and the termination codon decoded. The best studied examples of how these tRNAs are utilised has come from viruses in which translation of a specific termination codon results in the synthesis of a new protein containing the entire sequence of the usual translation product plus an extended C-terminal sequence. In the retrovirus Murine Leukaemia Virus (MLV) the synthesis of the <u>gag-pol</u> fusion protein requires translation of the UAG termination codon at the end of the <u>gag</u> reading frame, an event that again requires the presence of an mRNA structure (a stem-loop) located 3' to the leaky termination codon. The UAG codon in this case is decoded by a tRNA$^{Gln}$.

Random termination readthrough by natural suppressor tRNAS is clearly in itself suppressed in normal cells since a random C-terminal extension of cellular proteins must undoubtedly be detrimental to cell growth and viability even though there is a marked tendency for most natural mRNA terminators to have a second in-frame termination codon immediately downstream.

Hopping. This is a rather unusual non-standard mRNA decoding event with only one well described example, namely the event that occurs during the translation of the mRNA encoding the phage T4 topoisomerase subunit gene 60 (Weiss et al., 1990). With almost 100% efficiency a 50 nucleotide mRNA sequence which separates codons 46 from codon 47 is by-passed by the ribosome. The mechanism involved and the universality or otherwise of this phenomenon remains to be seen.

## SUMMARY AND CONCLUSIONS

Protein synthesis is essentially error-free in spite of the plethora of molecular interactions that are required during mRNA decoding. The role played by the ribosome is a crucial one and central to the control of the error rate to a level which is acceptable to normal cellular function. That translation is not 100% error-free suggests that the energetic costs of ensuring zero errors is to high a burden for the cell to carry (Kurland and Ehrenberg, 1985). The remarkable recent discoveries that certain translational events that were once viewed generically as random errors can occur in a programmed mRNA-specific fashion, to an efficiency in some cases approaching 100% does show that most cellular systems have evolved an ability to exploit translational errors albeit to a limited extent. The question that still remains to be answered is 'How universal is the use of programmed translational errors particularly in complex multicellular organisms where genetic material is not in short supply?'.

## ACKNOWLEDGEMENTS

Work on translational fidelity in the authors laboratory has been supported by funds from the SERC and NATO.

## REFERENCES

Atkins, J.F., Weiss, R.B. and Gesteland, R.F. (1990) Ribosome gymnastics - degree of difficulty 9.5, style 10.0. Cell, 62: 413-423.

Atkins, J.F., Weiss, R.B., Thompson, S. and Gesteland, R.F. (1991) Towards a genetic dissection of the basis of triplet decoding and its natural subversion: programmed reading frame shifts and hops. Ann. Rev. Genet. 25: 201-208.

Böck, A., Forchhammer, K., Heider, J., Leinfelder, W., Samers, G., Veprak, B. and Zinoni, D. (1991) Selenocysteine: the 21st amino acid. Mol. Microbiol. 5: 515-520.

Cornut, B. and Willson, R.C. (1991) Measurement of translational accuracy in vivo: missense reporting using inactive enzyme mutants. Biochimie 73: 1567-1572.

Edelmann, P. and Gallant, J. (1977) Mistranslation in E. coli. Cell 10: 131-137.

Firoozan, M., Grant, C.M., Duarte, J. and Tuite, M.F. (1991) Quantitation of readthrough of termination codons in yeast using a novel gene fusion assay. Yeast 7: 173-183.

Fox, T.D. (1987) Natural variation in the genetic code. Ann. Rev. Genet. 21: 67-91.

Gorini, L. (1974) Streptomycin and misreading of the genetic code. In "Ribosomes" (eds. M. Nomura, A. Tissieres and P. Lengyel) Cold Spring Harbor Laboratory, N.Y. pp 791-803.

Hatfield, D., Lee, B.J., Smith, D.W.E. and Oroszlan, S. (1990) Role of nonsense, frameshift and missense suppressor tRNAs in mammalian cells. Prog. Cell. Subcell. Biol. 11: 115-145.

Hinnebusch, A.G. and Liebman, S.W. (1991) Protein synthesis and translational control in Saccharomyces cerevisiae. In "The Molecular Biology and Cellular Biology of the Yeast Saccharomyces: Volume 1" (eds. J.R. Broach, J.R. Pringle and E.W. Jones) Cold Spring Harbor Laboratory, New York, pp. 627-736.

Hughes, D. (1991) Error-prone EF-Tu reduces in vivo enzyme activity and cellular growth. Mol. Microbiol. 5: 623-630.

Kirkwood, T.B.L., Rosenberger, R.F. and Galas, D.J. (eds.) (1986) "Accuracy in Molecular Processes", Chapman and Hall, London (a series of excellent reviews).

Kurland, C.G. and Ehrenberg, M. (1985) Constraints on the accuracy of mRNA movement. Q. Rev. Biophys. 18: 423-450.

Kurland, C.G. and Ehrenberg, M. (1987) Growth-optimising accuracy of gene expression. Ann. Rev. Biophys. Chem. 16: 291-317.

Murgola, E.J. (1985) tRNA suppression and the code. Ann. Rev. Genet. 19: 57-80.

Normanly, J., Ogden, R.C., Horvath, S.J. and Abelson, J. (1986) Changing the identity of a tRNA. Nature 321: 213-219.

Parker, J. (1989) Errors and alternatives in reading the

universal genetic code. Microbiol Rev. <u>53</u>: 273-298.

Piepersberg, W., Geyl, D., Hummel, H. and Bock, A. (1980) Physiology and biochemistry of bacterial ribosomal mutants. <u>In</u> 'Genetics and Evolution of RNA Polymerases, tRNA and Ribosomes' (ed. S. Osawa <u>et al</u>.), Tokyo University Press, Tokyo, pp. 359-375.

Pollard, J.W. (1984) Application of two-dimensional polyacrylamide gel electrophoresis to studies of mistranslation in animal and bacterial cells. <u>In</u> "Two-Dimensional Gel Electrophoresis of Proteins" (ed. J.E. Celis) Academic Press, New York, pp363-395.

Sandbanken, M. and Culberson, M.R. (1988) Mutations in elongation factor EF-1α affecting the frequency of frameshifting and amino acid misincorporation in <u>S. cerevisiae</u>. Genetics <u>120</u>: 923-934.

Scorer, C.A., Carrier, M.J. and Rosenberger, R.F. (1991) Amino acid misincorporation during high-level expression of mouse epidermal growth factor in <u>Escherichia coli</u>. Nucleic Acids Res. <u>19</u>: 3511-3516.

Weiss, R.B., Huang, W.M. and Dunn, D.M. (1990) A nascent peptide is required for ribosomal by pass of the coding gap in bacteriophage T4 gene 60. Cell <u>62</u>: 117-126.

# ISOLATION AND CHARACTERIZATION OF EUKARYOTIC TRANSLATION INITIATION, ELONGATION AND TERMINATION FACTORS

H. Trachsel
Institute of Biochemistry
and Molecular Biology
University of Berne
Bühlstrasse 28
3012 Berne
Switzerland

## INITIATION FACTORS

### Structure

Initiation factors were first isolated from mammalian cells. Using a cell-free system containing salt-washed rat ribosomal subunits, rabbit globin mRNA, charged tRNAs, ATP, GTP and purified translation elongation factors, Schreier, Erni and Staehelin (Schreier et al., 1977) defined a number of proteins  from a rabbit ribosomal salt wash fraction, which were required to promote globin synthesis *in vitro*. These proteins were purified to homogeneity and shown to be required for the binding of initiator Met-tRNA and mRNA to ribosomes (Trachsel et al., 1977, for a review, see Hershey, 1991). These factors and factors isolated in other laboratories together with some of their properties are listed in Table 1.

Using a similar identification strategy, initiation factors were also isolated from other sources including wheat germ. From wheat germ the factors eIF-1A, 2, 3, 4A, 4B, 4E, 4F and 5 were purified (Browning and Ravel, 1990 and references therein). Factor eIF-4F was found to exist in two isoforms containing two subunits each of 220 and 24 kD resp. 80 and 28 kD. The wheat germ and mammalian factors have similar subunit composition but the subunits have different molecular weights and do not or only poorly substitute for each other in the reconstituted systems.

NATO ASI Series, Vol. H 97
Post-transcriptional Control of Gene Expression
Edited by Orna Resnekov and Alexander von Gabain
© Springer-Verlag Berlin Heidelberg 1996

**Table 1: Mammalian translation initiation factors**

| Factor | Other name | Function(s) | Mass | Subunits | Cloned |
|---|---|---|---|---|---|
| eIF-1 | | Stimulation of Met-tRNA and mRNA binding to 40S ribosomes | 15 kD | | no |
| eIF-1A | eIF-4C | Stimulation of Met-tRNA and mRNA binding to 40S ribosomes | 17 kD | | yes |
| eIF-2 | | Met-tRNA binding to 40S ribosomes | 130 kD | $\alpha$: 36<br>$\beta$: 38<br>$\gamma$: 55 | yes<br>yes<br>yes |
| eIF-2A | | AUG-dependent Met-tRNA binding to 40S ribosomes | 65 kD | | no |
| eIF-2B | GEF | GDP:GTP exchange on eIF-2 | 270 kD | $\alpha$ : 26<br>$\beta$ : 39<br>$\gamma$ : 58<br>$\delta$ : 67<br>$\varepsilon$: 82 | yes<br>yes<br>yes<br>yes<br>yes |
| eIF-2C | Co-eIF-2A | Stabilization of ternary complex | 94 kD | | no |
| eIF-3 | | Ribosome dissociation. Stabilization of ternary complex. Stimulation of mRNA binding. | 550 kD | $\alpha$ : 35<br>$\beta$ : 36<br>$\gamma$ : 40<br>$\delta$ : 44<br>$\varepsilon$ : 47<br>$\zeta$ : 67<br>$\eta$ : 115<br>$\theta$ : 170 | no<br>no<br>no<br>no<br>no<br>no<br>no<br>no |
| eIF-3A | eIF-6 | Ribosome dissociation | 25 kD | | no |
| eIF-4A | | mRNA binding. RNA helicase | 44 kD* | | yes |
| eIF-4B | | mRNA binding. RNA helicase | 80 kD | | yes |
| eIF-4E | CBP I | mRNA binding. Cap binding | 24 kD ** | | yes |
| eIF-4F | CBP II | mRNA binding. Cap recognition. RNA helicase | 270 kD | $\alpha$ : 25**<br>$\beta$ : 43**<br>$\gamma$ : 220 | yes<br>yes<br>yes |
| eIF-5 | | Ribosomal subunit joining | 49 kD | | yes |
| eIF-5A | eIF-4D | Ribosomal subunit joining. Formation of the first peptide bond | 17 kD | | yes |

More recently, initiation factors have also been isolated from the yeast *Saccharomyces cerevisiae* (for a review, see Müller and Trachsel, 1990)

This system offers the possibility to combine powerfull genetic approaches with biochemical methods. Accordingly, most factors were identified by genetic approaches. The factors found so far are eIF-2, 2B, 4A, 4E, 4F, 5, and 5A. Genetic approaches have also led to the discovery of factors in *Saccharomyces cerevisiae* which were so far not found in the mammalian system. These are the factors Prt1(80 kD)(Hanic-Joice et al., 1987; Keierleber et al., 1986), Sui1 (15 kD) and Ssl2 (95 kD)(Gulyas and Donahue, 1992; Yoon and Donahue, 1992).

Abundancies of initiation factors in eukaryotic cells were measured in HeLa cells, rabbit reticulocytes and wheat germ. They vary quite strongly from over 1% of total protein for eIF-4A and eIF-5A to less than 0.1% for eIF-4F. This translates into ribosome to factor ratios of 3-5 copies for the highly abundant factors eIF-4A and eIF-5A, 0.2-1 copy per ribosome for the middle abundant factors such as eIF-2, 3, and 4B and 0.01 to 0.1 copy per ribosome for the low abundant factors such as eIF-2B, 4F and 5. Ribosome concentrations in eukaryotic cells are in the order of 1 $\mu$M.

Cloning and sequencing of mammalian and yeast cDNA and genomic DNA encoding translation initiation factors revealed that eukaryotic factors are highly conserved. Amino acid sequence identities between mammalian and yeast initiation factors range from 33% (eIF-4E) to 65% (eIF-4A). The factors eIF-2, 4A, 4B, and 5A carry unusual sequence motifs or amino acid modifications. The $\beta$-subunit of eIF-2 contains in its N-terminal half three lysine blocks with 7-8 lysine residues each and in its C-terminal half a zinc finger motif (Pathak et al., 1988). Mutations in this sequence motif influence AUG recognition by initiation complexes in yeast (Donahue et al., 1988). Factor eIF-4A shares short sequence motifs including the motif DEAD with a large number of proteins, the DEAD protein family (Linder et al., 1989). Since eIF-4A is part of an RNA helicase (see below) it is assumed that members of this family are involved in RNA secondary structure melting in many biological reactions. Sequence analysis of the cDNA encoding mammalian eIF-4B (Milburn et al., 1990) revealed the presence of two RNA binding motifs, RNP1 and RNP2, previously found in other RNA binding proteins. Factor eIF-5A carries a unique modificaton at a lysine residue. Through attachement of an amino-butyl moiety from spermidine to this residue followed by its hydroxylation it is posttranslationally changed to hypusine. This modification is essential and was so far only found in eIF-5A (Schnier, 1990).

The ability of initiation factors to form multicomponent complexes is probably an important aspect of their function. The factors eIF-4E, and 4A are found in a complex with the polypeptide p220. The resulting eIF-4F complex was found to be associated with the multi-subunit factor eIF-3 in crude initiation factor preparations. It appears therefore very likely that initiation factors form giant multicomponent complexes during initiation of translation in eukaryotes.

Relatively little is known about the regulation of expression of initiation factor genes. The expression of eIF-2$\alpha$, eIF-4A and eIF-5A is coordinated with total protein synthesis and is regulated at the transcriptional or RNA processing level. Factor eIF-4A exists in two forms encoded by two different genes in mouse which are over 90% identical. The two forms are differentially expressed in different tissues, but the biological significance of this finding is not known (Nielsen and Trachsel, 1988).


## Function

The general function of initiation factors is to select an mRNA for translation and to position a ribosome at the AUG initiator codon such that the open reading frame of the mRNA can be translated into a polypeptide.

Studies in reconstituted systems allowed the assignment of rough functions to each factor in the pathway of translation initiation. Assays include partial reactions such as ternary complex eIF-2-GTP-Met-tRNA formation, ternary complex binding to 40S ribosomal subunits, factor binding to mRNA, mRNA binding to 40S ribosomal subunits, unwinding of RNA duplex structures, Met-puromycin formation, ribosomal subunit joining and 80S initiation complex formation. Single omissions of factors in these assays defined factor functions.

Based on data from a large number of such experiments, the knowledge about scanning of mRNA by the ribosome resulting from the work of

Kozak (for a review, see Kozak, 1991) and contributions from genetic experiments in yeast the following model of translation initiation and factor function can be proposed (Fig 1):

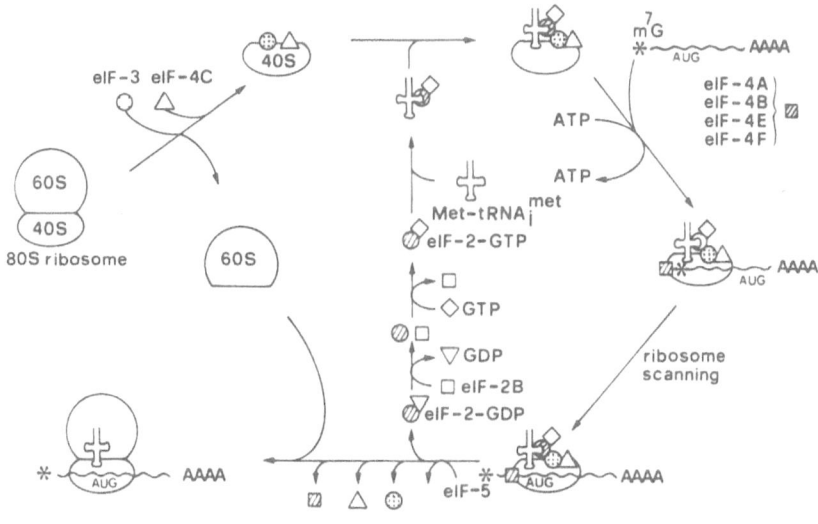

**80S initiation complex**

Fig. 1: Scheme of eukaryotic translation initiation. The cycle begins with the 80S ribosome and ends with the 80 initiation complex.
40S, small ribosomal subunit; 60S, large ribosomal subunit; eIF, eukaryotic initiation factor; $m^7G$, mRNA cap structure; met-tRNA$_i^{met}$, initiator methionyl transfer RNA.

80S ribosomes dissociate spontaneously into 40S and 60S ribosomal subunits at physiological $Mg^{2+}$ concentrations and are kept separated by binding of eIF-1A and eIF-3 to 40S subunits. In addition, ribosome dissociation factors such as eIF-3A may act on 80S ribosomes to dissociate them into subunits. 40S subunits bind the ternary complex eIF-2-GTP-Met-tRNA. This complex can form in the absence of other components but its formation may be stimulated by additional factors. The binding of the ternary complex to 40S ribosomes is stimulated by eIF-2C, eIF-3 and Prt1 (in yeast). The resulting 40S preinitiation complex binds to mRNA at or close to the 5' m7GpppX cap structure. The binding is catalyzed by eIF-4A, B, and F and requires hydrolysis of ATP. The factor eIF-4F recognizes and binds to the cap structure through its α-subunit, the cap binding factor eIF-4E. Alternatively, eIF-4E binding to the cap structure leads to binding of the β- and γ-subunits and assembly of eIF-4F at the cap structure. Binding of eIF-4B results in the formation of an active RNA helicase which unwinds the 5' region of the mRNA and allows 40S ribosome binding. The yeast factor Ssl2 may be involved in these reactions. The 40S ribosome then scans the mRNA and recognizes AUG codons and their flanking nucleotide sequences. The first AUG codon with appropriate flanking sequence is recognized as initiator AUG. Genetic experiments in yeast demonstrated that Met-tRNA, eIF-2 and Sui 1 fullfill important functions in this recognition process. Through the action of eIF-5 (and perhaps eIF-5A) GTP in the ternary complex is hydrolyzed to GDP, the factors are released from the 40S ribosome at the initiator AUG and the 60S ribosomal subunit joins the 40S to form an 80S ribosome. This 80S ribosome can begin translation of the open reading frame on the mRNA. Since eIF-2 has a 400-fold higher affinity for GDP than for GTP the exchange of GDP for GTP on eIF-2 under physiological concentrations of GDP and GTP requires the action of the exchange factor eIF-2B. This regenerates eIF-2 for a new round of initiation.

The rate-limiting step in the initiation pathway just described is the ATP-dependent binding of mRNA to ribosomes (Marcus, 1970).

More recently, it has become increasingly clear that the cap-dependent translation initiation pathway is not the only initiation pathway in eukaryotic cells. Translation of a number of viral and probably also cellular mRNAs is initiated internally. So far rather poorly characterized proteins can bind to internal ribosome entry sites and direct ribosomes to internal initiator AUGs (for a review, see Sonenberg, 1991).

Still another pathway of initiation is reinitiation at an AUG codon after translation of an upstream reading frame. The mechanism and factor requirements for this pathway remain to be determined (Dever et al., 1992).

## Open questions

The model described above hides the fact that the details of the whole pathway of initiation are rather poorly understood. A number of important questions remain to be answered: How do mRNA binding factors assemble on mRNA? How do mRNA binding factors unwind secondary structure in mRNA? Where does the ribosome bind on mRNA? Where does the 60S subunit join the 40S? Why are so many proteins involved in eukaryotes to direct the ribosome to the initiator AUG?

One might speculate that the involvement of a large number of factors catalyzing many steps in the pathway may offer the possibility to regulate at many steps. Indeed, regulation of translation through modification of translaton factor activity has been shown to occur at the step of ternary complex formation, ternary complex binding to the ribosome and mRNA binding to the ribosome (for a review, see Hershey, 1989).

## ELONGATION FACTORS

### Structure

Elongation factors were first isolated from rabbit reticulocytes more than a decade earlier than initiation factors. Since then they were purified from many eukaryotic organisms including human, rat, Xenopus, Drosophila, Artemia, Tomato and Saccharomyces cerevisiae. Furthermore, the genes encoding the factor eEF-1α from the sources mentioned were all cloned (for a review, see Riis and Merrick, 1990.
Six different factors, eEF-1α, β, γ, δ, eEF-2 and eEF-3 were identified. Some of their characteristics are described in Table 2.

Note, that the factor eEF-3 was only found in yeast and other fungi.
Since in both E. coli and eukaryotes certain UGA codons are translated into selenocysteine and the corresponding aminoacyl-tRNA is carried to the ribosome in E. coli by a special elongation factor (for a review, see Böck et al., 1991) it is reasonable to expect such an additional factor also in eukaryotes.

**Table 2: Mammalian translation elongation factors**

| Factor | Function(s) | Mass | Cloned |
|---|---|---|---|
| eEF-1α | Binding of aa-tRNA | 51 kD | yes |
| eEF-1β | GTP-GDP exchange on eEF-1α | 23 kD | yes |
| eEF-1γ | GTP-GDP exchange on eEF-1α | 49 kD | yes |
| eEF-1δ | GTP-GDP exchange on eEF-1α | 28 kD | no |
| eEF-2 | Translocation | 100 kD | yes |
| eEF-3 | Stimulation of aa-tRNA binding by eEF-1α | 116 kD | yes |

With up to 10 copies per ribosome or about 5% of total protein eEF-1α is the most abundant translation factor in eukaryotes. It is highly conserved among eukaryotes and the yeast factor is 30% identical to the E.coli factor EF-Tu. Two ore more genes encode eEF-1α in eukaryotes. Their expression is coordinated with the expression of ribosomal proteins and regulated at both the transcriptional and translational level. Like ribosomal proteins the mRNA encoding eEF-1α has a polypyrimidine stretch in the 5' untranslated region which acts as a binding site for a protein regulating mRNA translation. The factor is posttranscriptionally modified by methylation at lysine residues and by addition of ethanolamine-glycerylphosphate at two glutamic acid residues. The effects of these modifications on factor activity are not yet known.

A number of eEF-1α-like proteins have been found such as Xenopus 42Sp48 (Coppard and Clark, 1991), Drosophila F2 EF-1α, and yeast Sup 2 (Didichenko et al., 1991). Their functions remain to be elucidated. Very interestingly, overexpression of eEF-1α in Drosophila appears to lead to prolongation of the life span of this organism (Shepherd and Gehring, 1989).

The gene encoding eEF-1β has been cloned from Artemia only. The amino acid sequence shows no homology to other proteins except to eEF-1δ (from which only partial amino acid sequences are available). The gene encoding eEF-1γ was cloned from human, Xenopus and Artemia.

The sequences are not as highly conserved as the eEF-1α sequences. The identity between human and Xenopus is 75%, between Xenopus and Artemia only 58%.

The factor eEF-2 represents another strongly conserved translation factor in eukaryotes. It occurs in about one copy per ribosome in eukaryotic cells. Its abundance is closely correlated with the amount of ribosomes but the biochemical mechanisms underlying this control are not known. A single histidine residue in eEF-2 is modified postranscriptionally to diphtamide. At this unique structure eEF-2 can be modified by ADP-ribosylation through the action of toxins such as diphteria toxin. ADP-ribosylated factor is inactive in translation (Davydova and Ovchinnikov, 1990). Note that eEF-2 activity has also been shown to be regulated by phosphorylation (Ryazanov et al., 1988).

## Function

Elongation is a cyclic process resulting in the addition of one amino acid to the growing polypeptide chain per cycle. Factor eEF-1α forms a ternary complex with aminoacyl-tRNA and GTP. This complex binds to the A-site of the ribosome. In fungi this reaction is stimulated by eEF-3 and ATP. The role of eEF-3 in higher eukaryotes is assumed (but has never been shown) to be fullfilled by a ribosomal protein. Hydrolysis of the bound GTP to GDP results in the release of eEF-1α from the ribosome. The ribosomal peptidyl transferase center transfers the growing chain to the amino acid in the A-site. It has recently been shown for E. coli that the peptidyl transferase center is active in the absence of most if not all ribosomal proteins indicating that this enzymatic activity may be solely RNA (Noller et al., 1992). The exchange of GDP for GTP on eEF-1α is accelerated about 1000-fold by the action of eEF-1βγδ complex. The exchange reaction is catalyzed by the C-terminal domain of eEF-1β and eEF-1δ. eEF-1γ stimulates the reaction but cannot promote it by itself.
This recycles eEF-1α for a new round of elongation. The relative movement of mRNA and the ribosome by 3 nucleotides is induced by the binding of an eEF-2·GTP complex. GTP hydrolysis is not required for

translocation per se but facilitates the release of eEF-2 from the ribosome after translocation.

About 6-9 amino acids are added per second in eukaryotic cells, about half the speed of prokaryotic elongation. The accuracy of elongation is in the order of one misincorporation per $10^4$ amino acid incorporations. It is mediated by proofreading of amino acyl-tRNA synthetases and eEF-1α.

## TERMINATION FACTORS

### Structure

Only one termination factors has so far been isolated from eukaryotes and its cDNA cloned. It is a homodimer with a subunit molecular weight of 54 kD and was termed eRF. Very recently, it was found that tryptophanyl-tRNA synthetase is identical to eRF and that its expression is enhanced in interferon-treated cells (Buwitt et al., 1992). The functional significance of this finding is not clear at the moment.

### Function

The appearance of one of the three stop codons in the A-site of the ribosome leads to the binding of an eRF·GTP complex to the ribosome. The release factor induces the peptidyltransferase on the ribosome to hydrolyze the peptidyl-tRNA in the P-site. Hydrolysis of GTP mediates RF release from the ribosome. Finally, the ribosome is released from the mRNA.

## REFERENCES

Böck A, Forchhammer K, Heider J, Baron C (1991) Selenoprotein synthesis: an expansion of the genetic code. Trends Biochem Sci 16:463-467

Browning KS, Ravel JM (1990) Determination of the amounts of the protein synthesis initiation and elongation factors in wheat germ. J Biol Chem 265:17967-17973

Buwitt U, Flohr T, Bottger EC (1992) Molecular Cloning and Characterization of an Interferon Induced Human cDNA with Sequence Homology to a Mammalian Peptide Chain Release Factor. EMBO J 11(2):489-496

Coppard NJ, Clark FC (1991) 42Sp48 in previtellogenic Xenopus oocytes is structurally homologous to EF-1 alpha and may be a stage-specific elongation factor. J Cell Biol 112:237-243

Davydova EK, Ovchinnikov LP (1990) ADP-ribosylated elongation factor 2 (ADP-ribosyl-EF-2) is unable to promote translocation within the ribosome. Febs Lett 261:350-352

Dever TE, Feng L, Wek RC, Cigan AM, Donahue TF, Hinnebusch AG (1992) Phosphorylation of initiation factor 2α by protein kinase GCN2 mediates gene-specific translational control of GCN4 in yeast. Cell 68:585-596

Didichenko SA, Ter-Avanesyan M, Smirnov V (1991) Ribosome-bound EF-1 alpha-like protein of yeast Saccharomyces cerevisiae. Eur J Biochem 198:705-711

Donahue TF, Cigan AM, Pabich EK, Castillo Valavicius B (1988) Mutations at a Zn(II) finger motif in the yeast eIF-2β gene alter ribosomal start-site selection during the scanning process. Cell 54:621-632

Gulyas KD, Donahue TF (1992) SSL2, a suppressor of a stem-loop mutation in the HIS4 leader encodes the yeast homolog of human ERCC-3. Cell 69:1031-1042

Hanic-Joyce PJ, Singer RA, Johnston GC (1987) Molecular characterization of the yeast PRT1 gene in which mutations affect translation initiation and regulation of cell proliferation. J Biol Chem 262:2845-2851

Hershey JWB (1989) Protein phosphorylation controls translation rates. J Biol Chem 264:20823-20826

Hershey JWB (1991) Translational control in mammalian cells. Annu. Rev. Biochem. 60:717-755

Keierleber C, Wittekind M, Qin S, McLaughlin CS (1986) Isolation and characterization of PRT1, a gene required for the initiation of protein biosynthesis in Saccharomyces cerevisiae. Mol Cell Biol 6:4419-4424

Kozak M (1991) An analysis of vertebrate mRNA sequences: Intimations of translational control. J Cell Biol 115:887-903

Linder P, Lasko PF, Leroy P, Nielsen PJ, Nishi K, Schnier J, Slonimsky PP (1989) Birth of the D-E-A-D box. Nature 337:121-122

Marcus A (1970) Tobacco mosaic virus ribonucleic acid-dependent amino acid incorporation in a wheat embryo system. J Biol Chem 245:955-961

Milburn SC, Hershey JWB, Davies MV, Kelleher K, Kaufman RJ (1990) Cloning and expression of eukaryotic initiation factor 4B cDNA: sequence determination identifies a common RNA recognition motif. EMBO J 9:2783-2790

Mueller PP, Trachsel H (1990) Translation and regulation of translation in the yeast Saccharomyces cerevisiae. Eur J Biochem 191:257-261

Nielsen PJ, Trachsel H (1988) The mouse protein synthesis initiation factor 4A gene family includes two related functional genes which are differentially expressed. EMBO J 7:2097-2105

Noller HF, Hoffarth V, Zimniak L (1992) Unusual Resistance of Peptidyl Transferase to Protein Extraction Procedures. Science 256(5062):1416-1419

Pathak VK, Nielsen PJ, Trachsel H, Hershey JWB (1988) Structure of the β-subunit of translation initiation factor eIF-2. Cell 54:633-639

Riis B, Merrick WC (1990) Eukaryotic protein elongation factors. Trends Biochem Sci 15:420-424

Ryazanov AG, Shestakova EA, Natapov PG (1988) Phosphorylation of EF-2 by EF-2 kinase affects rate of translation. Nature 334:170-173

Schnier J (1991) Translation initiation factor 5A and its hypusine modification are essential for cell viability in the yeast S. cerevisiae. Mol Cell Biol 11:3105-3114

Schreier MH, Erni B, Staehelin T (1977) Initiation of mammalian protein synthesis. I. Purification and characterization of seven initiation factors. J Mol Biol 116:727-753

Shepherd JCW, Gehring WJ (1989) Fruit flies with additional expression of the elongation factor EF-1α live longer. Proc Natl Acad Sci USA 86:7520-7521

Sonenberg N (1991) Picornavirus RNA translation continues to surprise. Trends in Genetics 7:105-106

Trachsel H, Erni B, Schreier MH, Staehelin T (1977) Initiation of mammalian protein synthesis. II. The assembly of the initiation complex with purified initiation factors. J Mol Biol 116:755-767

Yoon H, Donahue TF (1992) The Sui1 suppressor locus in Saccharomyces cerevisiae encodes a translation factor that functions during tRNAmet recognition of the start codon. Mol Cell Biol 12:248-260

**The phosphorylation of eukaryotic initiation factor 2 and its role in the regulation of translation during viral infection.**

Simon R. Green and Michael B. Mathews
Cold Spring Harbor Laboratory
P.O. Box 100
Cold Spring Harbor
New York 11724
USA

With the increasing intensity of research into the processes of translation, it has become apparent that this phase of gene expression is often tightly regulated. Alteration of the phosphorylation state of an essential protein appears to be the predominant method that cells use to regulate biochemical activity, so it is no surprise that many of the steps in the translational pathway can be regulated by phosphorylation or dephosphorylation. Nevertheless, the linkage between such changes in covalent modification and modulation of enzymatic activity has been established *in vivo* for only three proteins (Hershey, 1990): eukaryotic elongation factor 2 (eEF-2), eukaryotic initiation factor 4E (eIF-4E), and eukaryotic initiation factor 2 (eIF-2). In mitosis eEF-2 is phosphorylated, resulting in an inhibition of translation because of an inability of the phosphorylated form of eEF-2 to catalyze the translocation of the peptidyl-tRNA during polypeptide chain elongation. The protein eIF-4E is an essential component of the cap binding complex, eIF-4F, which is responsible for bringing the 40S ribosomal subunit to the methylated cap of mRNA during translation. If eIF-4E is dephosphorylated, as occurs during heat shock and mitosis, the eIF-4F complex fails to deliver the ribosomal subunit to the mRNA. The best understood mode of translation control, however, involves the phosphorylation of the α subunit of eIF-2.

**Function of eIF-2.**

Three non-identical polypeptide chains, the α, β and γ subunits, associate to form eIF-2. Although both the α and β subunits can be phosphorylated *in vitro*, no functional significance has been established as yet for the phosphorylation of the β subunit. On the other hand, the cessation of translation resulting from the phosphorylation of the α subunit of eIF-2 (eIF-2α) is well characterized (Hershey, 1989; Hershey, 1990; Hershey, 1991). In one of the early steps of translational initiation, eIF-2 complexed with GTP is responsible for transporting the initiator Met-tRNA$_i$ to the 40S ribosomal subunit (reviewed in Moldave, 1985; Pain, 1986; Hershey, 1991). eIF-2 remains attached to the ribosome and has been implicated in translational start site selection during scanning of the mRNA, an event which occurs after the 40S ribosomal subunit has become associated with the mRNA through the

NATO ASI Series, Vol. H 97
Post-transcriptional Control of Gene Expression
Edited by Orna Resnekov and Alexander von Gabain
© Springer-Verlag Berlin Heidelberg 1996

action of eIF-4F. When the 60S ribosomal subunit binds to the small ribosomal subunit to complete the initiation phase, all the ribosome-bound initiation factors are released in a process that is catalyzed by eIF-5 and is dependent on the hydrolysis to GDP of the GTP complexed with eIF-2. Thus, eIF-2 is released from the ribosome in a binary complex with GDP. For eIF-2 to re-enter the initiation process, the GDP has to be exchanged for GTP. At physiological magnesium ion concentrations, eIF-2 has a much greater affinity for GDP than for GTP and the dissociation of the bound GDP is catalyzed by an additional factor termed the guanine nucleotide exchange factor (GEF or eIF-2B). When the $\alpha$ subunit of eIF-2 is phosphorylated on serine-51, eIF-2 can no longer be recycled by GEF because of the formation of a non-dissociable complex between eIF-2 and GEF. Consequently, a deficiency arises in the supply of eIF-2.GTP binary complexes needed for transporting the initiator tRNA to the ribosome (Proud, 1986). Because cells contain considerably less GEF than eIF-2, translation ceases when only a fraction of eIF-2$\alpha$ is phosphorylated.

## The eIF-2$\alpha$ kinases.

Three kinase have been identified which can phosphorylate eIF-2$\alpha$ on serine-51. These kinases are usually found in an inactive state and they are activated by different stimuli.

GCN2 kinase from the yeast *S. cerevisiae* is activated by uncharged tRNAs that accumulate during amino acid starvation, resulting in a general decrease in the rate of translation (Hinnebusch, 1990; Wek et al., 1990). Nevertheless, through an intricate mechanism, GCN2 kinase elicits an increase in the synthesis of the GCN4 gene product which in turn is responsible for the transcriptional activation of many of the genes involved in the amino acid biosynthetic pathway.

The heme controlled repressor (HCR) is found almost exclusively in reticulocytes and other cell types of the hemopoietic lineage (Jackson, 1991). In reticulocytes this kinase is activated by the absence of heme, thereby coupling the synthesis of globin with the availability of its prosthetic group. This kinase has been shown to be activated by a wide range of treatments *in vitro*, but the physiological significance of these stimuli remains to be ascertained.

The third kinase is the double-stranded (ds)RNA activated inhibitor of translation (DAI) which is ubiquitously distributed throughout mammalian cell types. As its name suggests, this kinase is activated by double-stranded RNA. It is an essential component of the interferon-induced host anti-viral response pathway, shutting down translation in an attempt to prevent viral proliferation (reviewed in Mathews, 1993). DAI has also been implicated in cellular differentiation (Petryshyn et al., 1984), the heat shock response (Dubois et al., 1989), transcriptional activation (Zinn et al., 1988) and, more recently, the inhibition of cell proliferation (Koromilas et al., 1992).

DAI is fastidious about the RNA that can cause its activation (Mathews, 1993). The enzyme is activated by dsRNA which is perfectly duplexed and over 30 bp in length. Furthermore, the dsRNA has to be at the correct concentration: 10-100 ng/ml of dsRNA activates DAI, but when the concentration is increased above this level it becomes inhibitory such that 1 μg/ml of dsRNA fails to activate the kinase. There appears to be no preferred binding sequence on dsRNA, but the overall topology of the RNA duplex is important. Thus, intercalation of ethidium bromide into dsRNA prevents activation of DAI, and, likewise, dsDNA duplexes and RNA:DNA hybrids are both unable to activate DAI (for references see Mathews, 1993). Activation of DAI is accompanied by autophosphorylation at multiple serine and threonine sites and it is in this activated state that DAI can phosphorylate eIF-2α. Two models have been proposed for the activation of DAI by dsRNA (see Figure 1). The first suggests that DAI contains one RNA binding site and that two molecules of DAI interact with a single RNA duplex, permitting activation through intermolecular autophosphorylation between the two DAI molecules. At dsRNA concentrations above the optimum, dimerization would not occur, thereby inhibiting intermolecular autophosphorylation (Kostura and Mathews, 1989). Alternatively, the second model proposes that DAI possesses two RNA binding sites, one of high affinity (the activation site) and one of low affinity (the inhibitory site). At suitable RNA concentrations, dsRNA binds to the activation site causing activation by intramolecular phosphorylation, but at higher RNA concentrations RNA also binds to the inhibitory site preventing intramolecular phosphorylation (Galabru et al., 1989).

**Viral countermeasures.**

Viruses and mammalian cells have interacted over a period of millions of years. These interactions have resulted in viruses evolving a plethora of countermeasures designed to prevent the activation of DAI, thereby maintaining viral proliferation and evading the host anti-viral response. The systems developed by viruses to thwart DAI activation are diverse and have been recently reviewed (Sonenberg, 1990; Mathews, 1993). They include:

(1)  controlling the levels of DAI available for activation, either by degradation of the protein itself (poliovirus) or by down-regulating the synthesis of DAI (human immunodeficiency virus-1);

(2)  the synthesis of viral proteins which are themselves dsRNA binding proteins, and sequester the activating dsRNA before it can complex with DAI (reovirus and vaccinia virus);

(3)  the activation of a cellular inhibitor of DAI (influenza virus);

(4)  the generation of a viral protein which mimics the eIF-2α subunit, thereby competing for the kinase active site (vaccinia virus);

(5)  inhibiting the activation process through the use of virus-derived RNAs (adeno-virus, Epstein Barr virus and HIV-1).

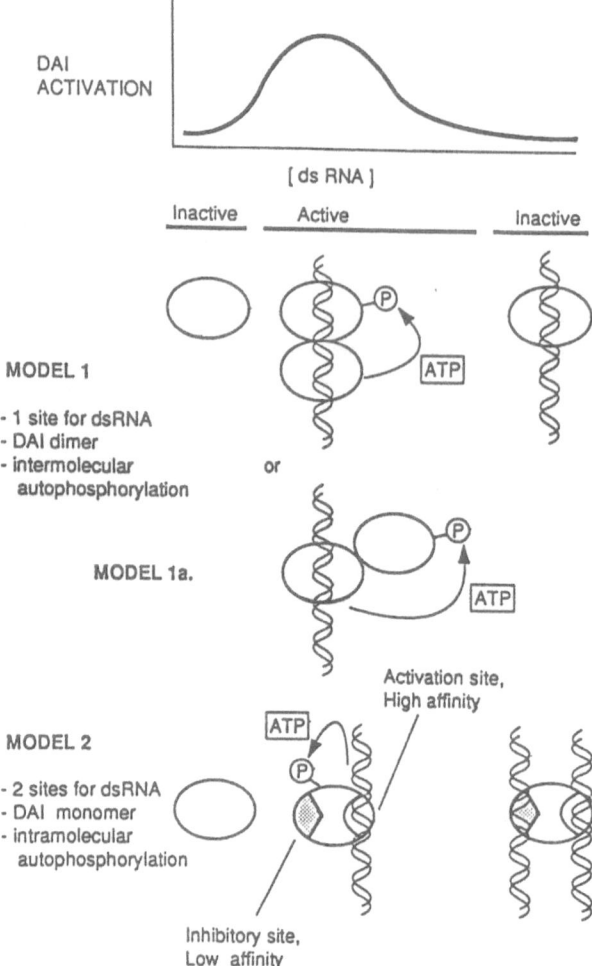

**Figure 1.** Models for the activation of DAI by dsRNA. Activation of DAI by dsRNA involves autophosphorylation and a suitable concentration of dsRNA (Mathews, 1993). Several models have been proposed to explain this phenomenon as illustrated here and discussed in the text.

The use of short RNA inhibitors was the first identified method used by viruses as a countermeasure for DAI activation and has been most intensively studied for adenovirus virus associated (VA) $RNA_I$. VA $RNA_I$ is an RNA polymerase III product that is 160 nucleotides long and is found predominantly in the cytoplasm of infected cells. Its synthesis begins during the early phase of infection and accelerates rapidly during the late phase of infection when it accumulates to over $10^8$ molecules per cell. The translational role of VA $RNA_I$ was recognized through the construction of a mutant virus (dl331) containing a disrupted VA $RNA_I$ gene. During the late phase of infection with wild type virus the cellular translational machinery is usurped so that only viral proteins are synthesized. At the same point during infection with dl331, however, little viral protein synthesis occurs. It was apparent that this was not a result of a failure to synthesize normal levels of functional viral mRNAs, implicating translation as the pathway in which VA $RNA_I$ functioned.

Further investigation confirmed the existence of a defect in translation and, moreover, showed that the cessation of translation was due to the activation of DAI (for references see Mathews and Shenk, 1991): (1) In a cell free extract derived from dl331 infected cells, translation was restored by the addition of either GEF or eIF-2; (2) the levels of phosphorylated eIF-$2\alpha$ and DAI were both increased in dl331 infected cells at late times of infection and there was a concomitant decrease in GEF activity; (3) infected cells contain virus derived dsRNA which probably arises from symmetrical transcription of the adenovirus genome; (4) the effect of the mutant virus was exacerbated by interferon, a known inducer of the kinase; (5) the mutant phenotype was suppressed by growth in cells which lack DAI or contain mutant forms of eIF-$2\alpha$ that are unable to be phosphorylated by DAI. The function of VA $RNA_I$ was confirmed by *in vitro* analysis which showed that VA $RNA_I$ can bind to DAI and inhibit its activation. In relation to the two previously discussed models (Figure 1), it is presumed that the high concentrations of VA $RNA_I$ that arise at the late stages of viral infection, prevent DAI activation by blocking dimerization (model 1) or by directly binding to the inhibitory site (model 2).

Experiments have been also performed *in vivo* with Epstein Barr virus EBER RNA and HIV-1 TAR RNA to determine whether these function to prevent DAI activation within the cell. EBER RNA can partially substitute for VA $RNA_I$ during dl331 infection (Bhat and Thimmappaya, 1985), while TAR RNA can rescue translation in a transient expression assay (Gunnery et al., 1992). The analogy to VA $RNA_I$ is not complete for either of these RNAs, however, and more extensive experiments need to be performed. Significantly, it is unlikely that either of these RNAs accumulates in the cell to a level equivalent to that of VA $RNA_I$ and necessary for the prevention of global DAI activation. One possibility is that these two RNAs may function by localized inhibition of DAI in regions of the cell in which viral mRNAs are concentrated.

**RNA-Protein Interactions.**

The recent cloning of a cDNA encoding the 551 amino acids of DAI (Meurs et al., 1990) permitted the use of molecular techniques to determine which amino acids in DAI interact with RNA. The C-terminal half of the protein contains the 11 catalytic subdomains essential for kinase activity (Hanks et al., 1988) (Figure 2). Within the N-terminal half of the protein there is a high percentage of basic amino acids that are clustered into three distinct regions (Figure 2). C-terminal truncation analysis was used to locate the RNA binding domain of DAI within the first 171 residues of the N-terminus (Katze et al., 1991; Patel and Sen, 1992; McCormack et al., 1992; Green and Mathews, 1992). This domain encompasses the first two regions of basic amino acids, both of which include one copy of a novel RNA binding motif which computer homology studies have also located in a number of other RNA binding proteins including TAR RNA binding protein, the mouse DAI homologue, vaccinia virus E3L protein, rotavirus NS34 protein, *E. coli* RNase III and the Drosophila protein Staufen (Green and Mathews, 1992; St Johnston et al., 1992). There are regions of similarity across the entire 67 residues of this motif but the C-terminal portion is especially well conserved (Figure 2). Furthermore, in all copies of this motif so far identified, an $\alpha$-helix is predicted to be present within this area of highest homology (Green and Mathews, 1992; McCormack et al., 1992).

Within the RNA binding domain of DAI, the tandem repeat of this motif is essential, since the deletion of either copy prevents the association between DAI and dsRNA or VA RNA$_I$. However, mutations distributed throughout the RNA binding domain indicate that the first copy of the motif is more important for RNA binding than the second. In particular, mutations which altered either the net positive charge or the structure of the predicted $\alpha$-helix within the first motif completely abrogated the ability of DAI to bind RNA. The dominance of motif 1 was confirmed by the observation that a mutant containing a duplication of motif 1 was still capable of binding RNA while the equivalent mutant containing a tandem repeat of motif 2 failed to bind RNA. Thus, although a tandem repeat of this motif is required for RNA binding, the first copy of the motif makes the more important and the majority of the protein:RNA interactions (Green and Mathews, 1992).

The size of dsRNA determines its ability to activate DAI. Duplexes longer than 30 bp bind and activate the enzyme with an efficiency that increases with increasing chain length, reaching an optimum at around 85 bp (Manche et al., 1992). Molecules shorter than 30 bp fail to bind stably and do not activate the kinase, but, like longer duplexes, inhibit DAI activation at high RNA concentrations. The differing affinities of DAI for these RNAs is not a result of different amino acids being responsible for binding different lengths of dsRNA, since mutations within DAI have the same effect on RNA binding regardless of chain length (Green and Mathews, 1992). A possible explanation for this phenomenon draws an analogy to some DNA binding proteins such as zinc finger proteins which interact with DNA through the inser-

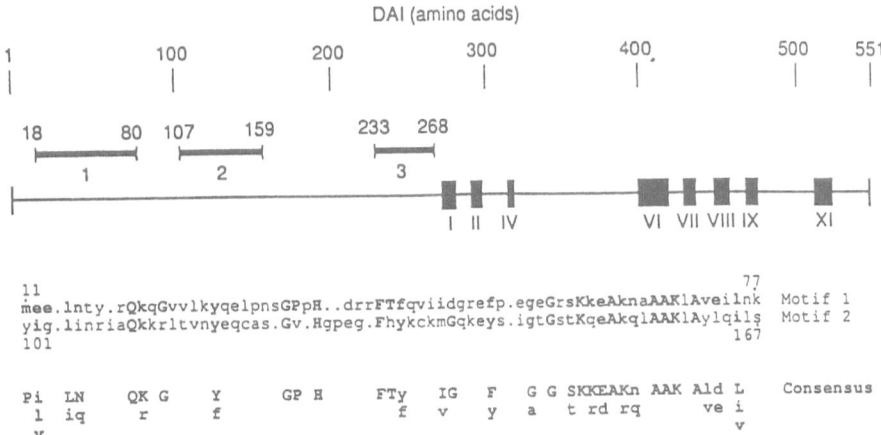

**Figure 2. Structure of DAI.** Map of DAI (top) showing the basic regions (1-3), and kinase subdomains (I-XI). RNA binding motifs from human DAI and consensus sequence from human and mouse DAI, vaccinia E3L, rotavirus NS34, TAR RNA binding protein and staufen (bottom). Similar or identical amino acids present in the same position in four or more of the sequences are shown in bold type as lower or upper case letters respectively.

tion of an α-helix into the major groove. In this analogy, the α-helix in each copy of the RNA binding motif within DAI is inserted into the minor groove of the dsRNA helix. The minor groove would be used because it is wide and shallow in the A-form configuration of RNA, while the major groove is deep but too narrow to permit effective interactions. If the dsRNA is of optimal length, then both α-helices could interact with the RNA in such a manner as to cause a conformational change in the protein (such as the removal of a pseudosubstrate site from active site of the kinase) thereby permitting autophosphorylation and activation. As the length of the RNA duplex is decreased, the probability of the two motifs interacting with the dsRNA is also reduced, explaining why shorter dsRNAs have a lesser ability to interact with and activate DAI. RNA duplexes below 30 bp are presumably too small to permit binding of both α-helices, so they do not support the conformational change and are unable to activate DAI. However, we presume that they can still interact weakly with the first copy of the motif, explaining how they act as inhibitors at high concentrations (Green and Mathews, 1992).

VA RNA$_I$ appears to interact with the same amino acids in DAI as dsRNA, yet it does not activate DAI. A possible interpretation of this is that although the two motifs interact with VA RNA$_I$, the structure of this RNA ensures that the interaction occurs in such a manner to prevent the conformational change required for activation. The fact that VA RNA$_I$ and dsRNA appear to bind to the same site on DAI would support the idea that activation is dependent on a single RNA binding site and intermolecular phosphorylation (model 1). Additional evidence to further support the intermolecular model was generated using truncated forms of DAI. The first truncation was a naturally occurring degradation product (p48) which does not possess the RNA binding domain, while the second truncation removed all the domains necessary for kinase function and contained only the N-terminal RNA binding domain. Neither of the truncated proteins was able to autophosphorylate and activate themselves, but both of them acted as a substrate for activated wild type DAI, indicating that intermolecular phosphorylation is occurring (unpublished results). These data would suggest that a modification of the first model is required (Figure 1, model 1a). Although dimerization is essential, it need not necessarily be between two DAI molecules attached to the same dsRNA duplex; rather, it could occur when one DAI molecule is bound to dsRNA and the other is unbound. In this version of the first model, high concentrations of dsRNA or VA RNA$_I$ would inhibit DAI activation because all the DAI would be complexed with RNA and therefore none would remain unbound and available for dimerization.

To summarize, cells have developed an elaborate pathway by which they limit viral infection. The protein kinase DAI is the pivotal agent within this process. DAI is activated during viral infection, causing a cessation of translation seemingly in an attempt by the cell to commit suicide, or at least suspend vital activities for a period, thus preventing further virus proliferation. In response viruses have developed a series of countermeasures aimed at disrupting the functioning of DAI, and thereby neutralizing the host anti-viral defense mechanism.

## Acknowledgements

We thank P. Clarke and S. Gunnery for helpful discussions. Work in our laboratory was supported by a National Cancer Institute program project grant (NIH CA13106).

## References

Bhat, R.A. and Thimmappaya, B. (1985). Construction and analysis of additional adenovirus substitution mutants confirm the complementation of VAI RNA function by two small RNAs encoded by Epstein-Barr virus. J. Virol. 56, 750-756.

Dubois, M.-F., Galabru, J., Lebon, P., Safer, B., and Hovanessian, A.G. (1989). Reduced activity of the interferon-induced double-stranded RNA-dependent protein kinase during a heat shock stress. J. Biol. Chem. 264, 12165-12171.

Galabru, J., Katze, M.G., Robert, N., and Hovanessian, A.G. (1989). The binding of double-stranded RNA and adenovirus VAI RNA to the interferon-induced protein kinase. Eur. J. Biochem. 178, 581-589.

Green, S.R. and Mathews, M.B. (1992). Two RNA binding motifs in the double-stranded RNA activated protein kinase, DAI. Genes Dev. 6, 2478 2490.

Gunnery, S., Green, S.R., and Mathews, M.B. (1992). HIV-1 TAR RNA stimulates protein synthesis in vivo and in vitro: Relationship between structure and function. Proc. Natl.

Acad. Sci. USA *89*, 11557-11561.

Hanks, S.K., Quinn, A.M., and Hunter, T. (1988). The protein kinase family: conserved features and deduced phylogeny of the catalytic domains. Science *241*, 42-52.

Hershey, J.W.B. (1989). Protein phosphorylation controls translation rates. J. Biol. Chem. *264*, 20823-20826.

Hershey, J.W.B. (1990). Overview: Phosphorylation and Translational Control. In Translationally regulated genes in higher eukaryotes. R.E. Thach, ed. (Basel: S. Karger), pp. 17-27.

Hershey, J.W.B. (1991). Translational control in mammalian cells. Annu. Rev. Biochem. *60*, 717-755.

Hinnebusch, A.G. (1990). Involvement of an initiation factor and protein phosphorylation in translational control of *GCN4* mRNA. TIBS *15*, 148-152.

Jackson, R.J. (1991). Binding of Met-tRNA. In Translation in eukaryotes. H. Trachsel, ed. (Boca Raton, USA: CRC Press), pp. 193-230.

Katze, M.G., Wambach, M., Wong, M.-L., Garfinkel, M., Meurs, E., Chong, K., Williams, B.R.G., Hovanessian, A.G., and Barber, G.N. (1991). Functional expression and RNA binding analysis of the interferon-induced, double-stranded RNA-activated, 68,000-M$_r$ protein kinase in a cell-free system. Mol. Cell. Biol. *11*, 5497-5505.

Koromilas, A.E., Roy, S., Barber, G.N., Katze, M.G., and Sonenberg, N. (1992). Malignant transformation by a mutant of the IFN-inducible dsRNA-dependent protein kinase. Science *257*, 1685-1689.

Kostura, M. and Mathews, M.B. (1989). Purification and activation of the double-stranded RNA-dependent eIF-2 kinase DAI. Mol. Cell. Biol. *9*, 1576-1586.

Manche, L., Green, S.R., Schmedt, C., and Mathews, M.B. (1992). Interactions between double-stranded RNA regulators and the protein kinase DAI. Mol. Cell. Biol. *12*, 5238-5248.

Mathews, M.B. (1993). Seminars in Virology (Philadelphia: Saunders).

Mathews, M.B. and Shenk, T. (1991). Adenovirus virus-associated RNA and translational control. J. Virol. *65*, 5657-5662.

McCormack, S.J., Thomis, D.C., and Samuel, C.E. (1992). Mechanism of interferon action: Identification of a RNA binding domain within the N-terminal region of the human RNA-dependent P1/eIF-2α protein kinase. Virology *188*, 47-56.

Meurs, E., Chong, K., Galabru, J., Thomas, N.S.B., Kerr, I.M., Williams, B.R.G., and Hovanessian, A.G. (1990). Molecular cloning and characterization of the human double-stranded RNA-activated protein kinase induced by interferon. Cell *62*, 379-390.

Moldave, K. (1985). Eukaryotic protein synthesis. Ann. Rev. Biochem. *54*, 1109-1149.

Pain, V.M. (1986). Initiation of protein synthesis. Biochem. J. *235*, 625-637.

Patel, R.C. and Sen, G.C. (1992). Identification of the double-stranded RNA-binding domain of the human interferon-inducible protein kinase. J. Biol. Chem. *267*, 7671-7676.

Petryshyn, R., Chen, J.-J., and London, I.M. (1984). Growth-related expression of a double-stranded RNA-dependent protein kinase in 3T3 cells. J. Biol. Chem. *259*, 14736-14742.

Proud, C.G. (1986). Guanine nucleotides, protein phosphorylation and the control of translation. Trends in Biochem. Sci. *11*, 73-77.

Sonenberg, N. (1990). Measures and countermeasures in the modulation of initiation factor activities by viruses. New Biol. *2*, 402-409.

Wek, R.C., Ramirez, M., Jackson, B.M., and Hinnebusch, A.G. (1990). Identification of positive-acting domains in GCN2 protein kinase required for translational activation of *GCN4* expression. Mol. Cell. Biol. *10*, 2820-2831.

Zinn, K., Keller, A., Whitemore, L.-A., and Maniatis, T. (1988). 2-aminopurine selectively inhibits the induction of β-interferon, c-*fos*, c-*myc* gene expression. Science *240*, 210-213.

# SELECTIVE PROTEOLYSIS BY THE UBIQUITIN SYSTEM

Stefan Jentsch[*], Ruth Heinlein, Joern Jungmann, Hans-Albert Reins,
Stephan Schlenker, Wolfgang Seufert, Thomas Sommer, and Sebastian Springer

Friedrich-Miescher-Laboratorium der Max-Planck-Gesellschaft
Spemannstr. 37-39, 7400 Tübingen, Germany. [*]Corresponding author

## Introduction

Intracellular protein levels are controlled by protein synthesis and degradation. A major proteolytic pathway of eukaryotes is ATP-dependent and requires the covalent attachment of ubiquitin, a small and highly conserved protein, to proteolytic substrates prior to degradation (for reviews see Finley & Chau 1991, Hershko & Ciechanover 1992, and Jentsch 1992a and b). This pathway is highly selective and mediates the elimination of abnormal proteins and controls the half-lives of some regulatory proteins. Known targets include transcriptional regulators (Hochstrasser et al. 1991), p53 (Scheffner et al. 1990), the Mos kinase (Nishizawa et al. 1992) and cyclins (Glotzer et al. 1991).

The conjugation of ubiquitin to protein substrates is a multistep process. In an initial activating step, a thioester is formed between the C-terminus of ubiquitin and an internal cysteine residue of a ubiquitin-activating (E1) enzyme. Activated ubiquitin is then transferred to a specific cysteine residue of one of several ubiquitin-conjugating (E2) enzymes. Finally, these E2 enzymes (sometimes in conjunction with specific substrate recognition proteins, E3s) donate ubiquitin to protein substrates. This results in branched protein conjugates in which the C-terminus of ubiquitin in linked by an isopeptide bond to specific internal lysine residues of target proteins. Since ubiquitin itself is a substrate for ubiquitination, certain substrates are modified by the attachment of multiubiquitin chains (Chau et al. 1989). Multiubiquitinated substrates are degraded *in vivo* by the proteasome, a multisubunit protease complex (Seufert & Jentsch, 1992).

The gene for ubiquitin-activating enzyme, *UBA1*, has been cloned from the yeast *S. cerevisiae* and shown to be essential for viability (McGrath et al. 1991). Ten genes coding for yeast ubiquitin-conjugating enzymes (*UBC* genes) have been isolated so far (reviewed in Jentsch 1992a and b). These enzymes are located in the cytosol and the nucleus, indicating that both cellular compartments have active ubiquitin-conjugating systems. In this extended abstract we summarize recent genetic data on the cellular functions of yeast ubiquitin-conjugating enzymes.

NATO ASI Series, Vol. H 97
Post-transcriptional Control of Gene Expression
Edited by Orna Resnekov and Alexander von Gabain
© Springer-Verlag Berlin Heidelberg 1996

**Cellular Functions of Ubiquitin-Conjugating Enzymes**

*Stress response*

Ubiquitin-dependent protein degradation is an essential function of the eukaryotic stress response. Polyubiquitin genes from various organisms are stress-inducible and the corresponding gene from yeast, *UBI4*, is required for viability under stress conditions (Finley et al. 1987). Similarly, the transcription of *UBC4* and *UBC5*, encoding two yeast ubiquitin-conjugating enzymes, is heat shock-inducible and the *ubc4 ubc5* double mutant is inviable under stress conditions (Seufert & Jentsch 1990). The *UBC1* encoded protein is closely related to UBC4 and UBC5. *UBC1*, *UBC4* and *UBC5* constitute a gene family essential for viability (the triple mutant is inviable) (Seufert & Jentsch 1990, Seufert et al. 1990). *ubc1* mutants exhibit a specific defect in growth after germination of ascospores, suggesting that UBC1 may have an additional function during this transition period of the yeast life cycle (Seufert et al. 1990). Structurally close homologs of *UBC4* and *UBC5* have been cloned from *Drosophila* (Treier et al. 1992) and *Caenorhabditis elegans* (Zhen et al. 1993). These genes can functionally replace yeast *UBC4*, suggesting that UBC4-like enzymes mediate similar functions in probably all eukaryotes.

*Cadmium resistance*

*UBC7*, encoding another yeast ubiquitin-conjugating enzyme, was shown to be neither important under normal growth conditions nor under heat-shock conditions (Jungmann et al. 1993). Interestingly however, *ubc7* mutants are hypersensitive to the heavy metal cadmium (not copper, not zinc). Similar cadmium sensitivities were also observed with *ubc1* and *ubc4* mutants. Moreover, mutants in the proteasome that degrades ubiquitin-protein conjugates are also hypersensitive to cadmium, suggesting that cadmium resistance is mediated in part by degradation of abnormal proteins (Jungmann et al. 1993). This implies that a major reason for cadmium toxicity may be cadmium-induced formation of abnormal proteins.

*Protein degradation at the endoplasmic reticulum*

The UBC6 protein has a C-terminal signal-anchor sequence and is an integral membrane protein of the endoplasmic reticulum, with the active site of the enzyme on the cytoplasmic face of the membrane (T. Sommer & S. Jentsch, in prep.). Recent genetic data indicate that

UBC6 may have a function in degradation of membrane proteins of the ER ("ER-degradation").

## Peroxisome biogenesis

A genetic screen in yeast identified several genes required for peroxisome biogenesis (Erdmann et al. 1989). One of these genes, *PAS2/UBC10*, turned out to encode a ubiquitin-conjugating enzyme (Wiebel & Kunau 1992). Interestingly, UBC10 was found to be associated with peroxisomes, suggesting that the crucial targets of this enzyme are also peroxisomal proteins.

## DNA repair

The UBC2 protein (also known as RAD6) plays a prominent role in a central DNA-repair pathway of yeast (Jentsch et al. 1987). Mutants of *UBC2* are highly vulnerable to DNA-damaging agents and deficient in induced mutagenesis. Moreover, *UBC2* is required for sporulation and repression of retrotransposition in yeast (Picologlou et al. 1990). This regulatory role suggests that UBC2 might mediate the degradation of regulatory proteins such as transcriptional repressors for genes encoding proteins for DNA repair, sporulation and transposition functions. Alternatively, since this enzyme conjugates ubiquitin to histones *in vitro*, UBC2 might modulate chromatin structure (Jentsch et al. 1987). Functional homologs of *UBC2* have been found in *Schizosaccharomyces pombe* (Reynolds et al. 1990), *Drosophila* (Koken et al. 1991a), and man (Schneider et al. 1990, Koken et al. 1991b).

## Cell cycle control

Only two of the known ten *UBC* genes, *UBC3* and *UBC9*, are essential for viability and both of these encode nuclear enzymes that are required for cell cycle progression (Goebl et al. 1988, W. Seufert & S. Jentsch, unpublished). Whereas *UBC9* is required at $G_2/M$ phase of the cell cycle (Seufert & Jentsch, in prep.), *UBC3* (also known as *CDC34*) is essential for the transition from the $G_1$ into the S-phase of the cell cycle (Goebl et al. 1988). Temperature-sensitive *ubc3* mutants arrest at the restrictive temperature at $G_1$ with multiple elongated buds and have unreplicated DNA. The spindle pole bodies duplicate but fail to undergo separation. The *in vivo* substrates of the UBC3 and UBC9 enzymes are presently unknown but they may include cyclins.

## Acknowledgement

Research in this laboratory is supported by the Deutsche Forschungsgemeinschaft, Fonds der chemischen Industrie, the German-Israeli Foundation for Scientific Research and Development, and the Human Frontier Science Program.

## References

Chau, V., Tobias, J.W., Bachmair, A., Marriott, D., Ecker, D.J., Gonda, D.K., & Varshavsky, A. (1989) A multiubiquitin chain is confined to specific lysine in a targeted short-lived protein. Science 243:1576-1583

Erdmann, R., Veenhuis, M., Mertens, D., & Kunau, W.-H. (1989) Isolation of peroxisome-deficient mutants of *Saccharomyces cerevisiae*. Proc. Natl. Acad. Sci. 86:5419-5423

Finley, D. & Chau, V. (1991) Ubiquitination. Annu. Rev. Cell Biol. 7:25-69

Finley, D., Özkaynak, E., & Varshavsky, A. (1987) The yeast polyubiquitin gene is essential for resistance to high temperature, starvation, and other stresses. Cell 48:1035-1046

Glotzer, M., Murray, A.W., & Kirschner, M.W. (1991) Cyclin is degraded by the ubiquitin pathway. Nature 349:132-138

Goebl, M.G., Yochem, J., Jentsch, S., McGrath, J.P., Varshavsky A., & Byers, B. (1988) The yeast cell cycle gene *CDC34* encodes a ubiquitin-conjugating enzyme. Science 241:1331-1335

Hershko, A. & Ciechanover, A. (1992) The ubiquitin system for protein degradation. Annu. Rev. Biochem. 61:761-807

Hochstrasser, M., Ellison, M.J., Chau, V., & Varshavsky, A. (1991) The short-lived MATα2 transcriptional regulator is ubiquitinated *in vivo*. Proc. Natl. Acad. Sci. USA 88:4606-4610

Jentsch, S. (1992a) Ubiquitin-dependent protein degradation: a cellular perspective. Trends Cell Biol. 2:98-103

Jentsch, S. (1992b) The ubiquitin-conjugation system. Annu. Rev. Genet. 26:177-205

Jentsch, S., McGrath, J.P., & Varshavsky, A. (1987) The DNA repair gene *RAD6* encodes a ubiquitin-conjugating enzyme. Nature 329:131-134

Jungmann, J., Reins, H.-A., Schobert, C., & Jentsch, S. (1993) Resistance to cadmium mediated by ubiquitin-dependent proteolysis. Nature, 361, in press

Koken, M., Reynolds, P., Bootsma, D., Hoeijmakers, J., Prakash, S., & Prakash, L. (1991a) *Dhr6*, a *Drosophila* homolog of the yeast DNA-repair gene *RAD6*. Proc. Natl. Acad. Sci. 88:3832-3836

Koken, M., Reynolds, P., Jasperdekker, I., Prakash, S, Prakash, L., & Bootsma, D. (1991b) Structural and functional conservation of two human homologs of the yeast DNA-repair gene RAD6. Proc. Natl. Acad. Sci. 88:8865-8869

McGrath, J. P., Jentsch, S., & Varshavsky, A. (1991) *UBA1*: an essential yeast gene encoding ubiquitin-activating enzyme. EMBO J. 10:227-236

Nishizawa, M., Okazaki, K., Furuno, N., Watanabe, N., & Sagata, N. (1992) The "second codon rule" and autophosphorylation govern stability of Mos during meiotic cell cycle in *Xenopus* oocytes. EMBO J. 11:2433-2446

Picologlou, S., Brown, N., & Liebman, S.W. (1990) Mutations in *RAD6*, a yeast gene encoding a ubiquitin-conjugating enzyme, stimulate retrotransposition. Mol. Cell. Biol. 10:1017-1022

Reynolds, S., Koken, M., Hoeijmakers, J., Prakash, S., & Prakash, L. (1990) The *rhp6+* gene of *Schizosaccharomyces pombe*: a structural and functional homolog of the *RAD6* gene from the distantly related yeast *Saccharomyces cerevisiae*. EMBO J. 9:1423-1430

Scheffner, M., Werness, B.A., Huibregtse, J.M., Levine, A.J., & Howley, P.M. (1990) The E6 oncoprotein encoded by human papillomavirus types 16 and 18 promotes the degradation of p53. Cell 63:1129-1136

Schneider, R., Eckerskorn, C., Lottspeich, F., & Schweiger, M. (1990) The human ubiquitin carrier protein E2($M_r$=17 000) is homologous to the yeast DNA repair gene *RAD6*. EMBO J. 9:1431-1435

Seufert, W., & Jentsch, S. (1990) Ubiquitin-conjugating enzymes UBC4 and UBC5 mediate selective degradation of short-lived and abnormal proteins. EMBO J. 9:543-550

Seufert, W., & Jentsch, S. (1992) *In vivo* function of the proteasome in the ubiquitin pathway. EMBO J. 11:3077-3080

Seufert, W., McGrath, J.P., & Jentsch, S. (1990) *UBC1* encodes a novel member of an essential subfamily of yeast ubiquitin-conjugating enzymes involved in protein degradation. EMBO J. 9:4535-4541

Treier, M., Seufert, W., & Jentsch, S. (1992) *Drosophila UbcD1* encodes a highly conserved ubiquitin-conjugating enzyme involved in selective protein degradation. EMBO J. 11:367-372

Wiebel, F.F. & Kunau, W.-H. (1992) The Pas2 protein essential for peroxisome biogenesis is related to ubiquitin-conjugating enzymes. Nature 359:73-76

Zhen, M., Heinlein, R., Jones, D., Jentsch, S., & Candido, E.P.M. (1993) The *ubc-2* gene of *Caenorhabditis elegans* encodes a ubiquitin-conjugating enzyme involved in selective protein degradation. Mol. Cell. Biol. 13, in press

# THE ROLE OF MOLECULAR CHAPERONES HSP70 AND HSP60 IN PROTEIN FOLDING

Franz-Ulrich Hartl
Program of Cellular Biochemistry & Biophysics
Sloan-Kettering Institute
1275 York Avenue
New York, NY 10021
USA

## The problem of protein folding in vivo

Protein folding is a problem of fundamental biological importance. It has been known for more than three decades that all the information required for the acqisition of the native state is contained in the linear amino acid sequence of the polypeptide chain. Proteins can fold spontaneously *in vitro*, at least under carefully chosen conditions, and this has led to the view that also within cells newly-synthesized polypeptides reach their native state in an essentially spontaneous reaction. Only very recently has it been realized that this is not the case. Cells contain a complex machinery of proteins, folding catalysts and molecular chaperones, which mediate folding in the cytosol as well as within subcellular compartments such as mitochondria, chloroplasts and the endoplasmic reticulum (Hartl et al., 1992). Molecular chaperones, mostly constitutively expressed stress proteins, play a preeminent role in these processes.

In light of the difficulties often encountered when the renaturation of unfolded proteins is attempted *in vitro*, it is noteworthy that folding in the cell occurs with remarkably high efficiency. Generally, unfolded polypeptides have the tendency to misfold and aggregate, and this property should be most pronounced in the cellular environment with its high concentration of total protein (~0.3 g/ml) and of newly-synthesized, folding chains (30-50 μM) (Zimmerman and Trach, 1991). Nevertheless, the yield of protein folding *in vivo* reaches almost 100% (Gething et al., 1986). This is due to the action of molecular chaperones, helper proteins which interact with non-native polypeptides and prevent unproductive off-pathway reactions during folding (Ellis, 1987; Rothman, 1989; Gething and Sambrook, 1992).

Essential functions in protein folding have been established for the members of the heat-shock protein 70 and heat-shock protein 60 families. Several lines of evidence suggest that there may be an ordered pathway of chaperone-mediated folding in which Hsp70s and Hsp60s, regulated by additional components, cooperate. It is believed that Hsp70s interact first with a nascent polypeptide chain emerging from the ribosome or at the trans-side of a membrane during translocation. This leads to the stabilization of an early folding intermediate, perhaps resembling the 'molten globule', which is subsequently transferred to Hsp60 for mediation of folding to the native state. Such a pathway has been described for proteins which fold following import into mitochondria and is likely to exist in the bacterial cytosol as well. In the following sections I will summarize the main properties which have been established for the sequential action of Hsp70 and Hsp60 in protein folding.

**Mitochondrial protein import as model system**

The first evidence that protein folding reactions *in vivo* require the ATP-dependent function of stress proteins was obtained through the analysis of mitochondrial protein import. Most mitochondrial proteins are encoded by nuclear genes and are synthesized as precursors in the cytosol. They have to be transported in an unfolded state from the cytosol across the mitochondrial membranes (Hartl and Neupert, 1990). Although the step of membrane translocation itself is not yet understood in detail, studying the folding of the newly-imported proteins has contributed considerably to our present understanding of chaperone functions. In particular, this experimental system served to establish the potential of Hsp70 and Hsp60 to act in a sequential folding pathway.

<u>Stabilisation of an unfolded state by Hsp70 in the cytosol</u>. Most mitochondrial precursor proteins appear to traverse the outer and inner mitochondrial membranes at so-called translocation contact sites in an extended conformation (Rassow et al., 1990). The unfolded state of precursor proteins has to be maintained in the cytosol as a prerequisite for efficient, post-translational translocation. Cytosolic Hsp70 and probably additional components such as homologues of the *E. coli* stress protein DnaJ participate in preventing premature folding (Deshaies et al., 1988). The majority of newly-synthesized proteins appear to interact with Hsp70 as nascent chains emerging from the ribosome (Beckman et al., 1990), perhaps with the exception of those proteins which undergo co-translational transport into the endoplasmic reticulum. Mitochondrial precursor proteins probably reach the receptors at the outer mitochondrial membrane in a complex with the chaperone (Figure 1). As specified below, the release of the Hsp70-bound protein or its transfer to another component in the chaperone cascade occurs in a tightly regulated process, depending on two other stress proteins which cooperate with Hsp70.

<u>Mitochondrial Hsp70 is required for translocation</u>. Following the binding of a mitochondrial precursor protein at specific receptor sites in the outer membrane, it is released from the cytosolic chaperone in an ATP-dependent manner and translocation into the matrix compartment proceeds through a hydrophilic environment, most likely a proteinaceous pore (Pfanner et al., 1987). Insertion of the positively charged N-terminal targeting sequence and its penetration into the inner membrane is driven by the membrane potential $\Delta Y$ (Martin et al., 1991a), while translocation of the mature protein part depends mainly on other driving forces. The mitochondrial form of Hsp70 (mHsp70) is of crucial importance for this step (Figure 1). In a yeast mutant defective in the function of mHsp70, precursor proteins

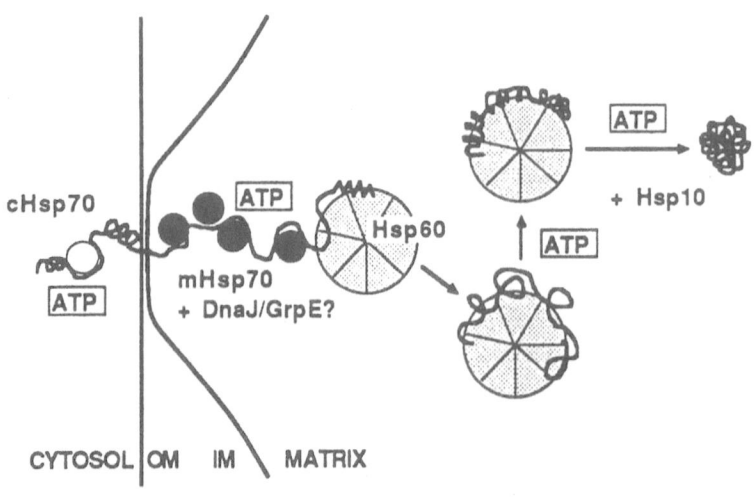

Figure 1. The pathway of chaperone-mediated protein folding in mitochondria

accumulate as intermediates spanning both membranes in a permanent association with mHsp70 (Kang et al., 1990). Binding of mHsp70 appears to be of importance in pulling the protein into the matrix in a tug of war with the cytosolic chaperones. This "pulling function" is defective in the mutant. If the unfolded state is a prerequisite for translocation, then import of proteins following their artificial unfolding in 8 M urea and the removal of cytosolic chaperones should be less affected in the mutant mitochondria. This is indeed the case. Precursor proteins which are diluted from denaturant into the import reaction are able to traverse the membranes of the mutant mitochondria, but then remain associated with the mutant mHsp70 as trapped folding intermediates. The interaction of a translocating polypeptide chain with the wild-type form of mHsp70 at the matrix side of the inner membrane (Scherer et al.

1990, Ostermann et al., 1990) may resemble that of cytosolic Hsp70 with nascent chains emerging from ribosomes. Import into mitochondria (and membrane translocation of proteins in general) is an unidirectional process. How this is achieved is unknown, but posttranslocational events such as proteolytic processing (and the glycosylation of secretory proteins), and the interaction with molecular chaperones following translocation are probably important in preventing a protein from slipping back into the cytosol.

<u>Transfer to Hsp60 for ATP-dependent folding</u>. Folding to the native state requires that the newly-imported proteins be transferred from mHsp70 to the chaperonin Hsp60 (Kang et al., 1990; Neupert et al., 1990; Manning-Krieg et al., 1991) (see Figure 1). Hsp60 mediates the folding of proteins to their native state or into folded monomers that subsequently undergo oligomeric assembly (Cheng et al., 1989). Hsp60-dependent folding requires ATP-hydrolysis (Ostermann et al., 1989) and the regulation of Hsp60 by Hsp10, which is the homologue of *E. coli* GroES (see below). The importance of Hsp60 for the biogenesis of mitochondrial proteins has again been established using a specific temperature sensitive yeast mutant. The defect in Hsp60 of the mutant strain *mif4* causes the misfolding of a variety of imported proteins such as the β subunit of the $F_1$-ATPase or the trimeric matrix enzyme ornithine transcarbamoylase (Cheng et al., 1989). These proteins aggregate upon import into the mutant mitochondria. The defect in Hsp60 affects also precursor proteins destined for the intermembrane space, such as cytochrome $b_2$, which are first imported into the matrix and then re-exported across the inner membrane (Hartl and Neupert, 1990). In the model of "conservative sorting", a hydrophobic signal-sequence mediates the re-export of these proteins into the intermembrane space (Glick et al., 1992). Presence of this hydrophobic sequence, either by interfering with folding of the mature protein part and/or by direct interaction with Hsp60, can result in a tight association of the protein with Hsp60 which serves to prevent folding prior to export (Koll et al., 1992).

# Mechanisms of chaperone action

Protein-mediated protein folding has been reconstituted *in vitro* with pure chaperone components (mostly of *E. coli*) and artificially unfolded substrate proteins. These studies have yielded insight into the molecular mechanisms of Hsp70 and Hsp60 action and into how these components may cooperate in the ordered protein folding pathway observed in mitochondria. It appears likely that homologues of two additional *E. coli* stress proteins, DnaJ and GrpE, are involved in this reaction. We can now define the distinct steps of this pathway:

<u>Stabilisation of an early folding intermediate</u>. Hsp70 family members consist of two domains, an N-terminal ATP-hydrolyzing domain of about 44 kD (Flaherty et al., 1990), and a C-terminal domain which is assumed to contain the substrate binding site. ATP-hydrolysis is generally required for the release of bound substrate proteins and the binding of peptides or unfolded proteins stimulates the ATP-ase of Hsp70 (Flynn et al., 1991). The *E. coli* homologue of Hsp70, DnaK, has been shown to interact with two other heat shock proteins, DnaJ and GrpE, by biochemical and genetic experiments. The genes encoding these three proteins were originally discovered through mutations blocking DNA-replication of bacteriophage lambda (Georgopoulos et al., 1990). DnaK, DnaJ and GrpE function in this process by disrupting a preprimsosmal protein complex at oriλ (Alfano and McMacken, 1989; Zylicz et al., 1989). This trio of stress proteins has also been shown more recently to drive the disassembly of RepA dimers into active monomers (Wickner, S. et al., 1991). The functional cooperation of DnaK, DnaJ, and GrpE in protein folding was revealed through studies using the monomeric mitochondrial protein rhodanese as a model substrate (Figure 2). When unfolded rhodanese is diluted from guanidnium-Cl into buffer solution, this results in rapid aggregation. The presence of DnaK can suppress this process, albeit with low efficiency, by formation of a complex between DnaK and an early folding intermediate of rhodanese (Langer et al., 1992a). DnaJ alone is also able to stabilize a folding

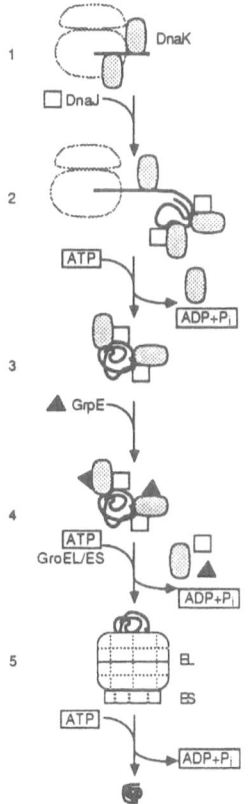

Figure 2. Model for the chaperone-mediated folding pathway of newly synthesized proteins in the bacterial cytosol. The reaction scheme corresponds to the steps involved in the folding of mitochondrial proteins (see Langer et al., 1992a).

intermediate of rhodanese against aggregation. Interestingly, both Dnak and DnaJ act synergistically in this process. A complex of rhodanese-DnaK-DnaJ is then formed which is not disrupted by ATP-hydrolysis but rather is stabilized by the presence of hydrolyzable ATP (Fig. 2, stage 3).

The synergistic effect of DnaK and DnaJ in stabilizing an early intermediate in protein folding may be due to both chaperones recognizing different conformational features of the polypeptide. Such differential recognition by DnaJ and DnaK is inferred from binding studies with defined folding intermediates of alpha-lactalbumin. Whereas DnaK is able to bind the carboxymethylated form of alpha-lactalbumin, which is extended and lacks stable secondary structure, DnaJ exhibits only a weak affinity for this state. Its binding affinity appears to increase with intermediates of alpha-lactalbumin which contain secondary structure and resemble the molten globule. The preference of DnaK for extended conformations has also been demonstrated analyzing the conformation of a Hsp70-bound peptide by nuclear magnetic resonance (Landry et al., 1992).

GrpE-dependent transfer to GroEL. Folding to the native state requires that the chaperone-bound intermediates are released from DnaK and DnaJ in a step which is mediated by the heat shock protein GrpE. The interaction with a preformed rhodanese-DnaK-DnaJ protein complex leads to the dissociation of the complex allowing the transfer of the folding intermediate to another chaperone such as the Hsp60 homologue of *E. coli*, GroEL. GrpE thus can be regarded as a coupling factor of the Hsp70 and Hsp60 systems. GrpE fulfills this function by directly interacting with DnaK and stimulating nucleotide exchange from an ADP- to an ATP-bound form of DnaK (Fig. 2, stage 4). A similar reaction may also be involved in the release of chaperone-bound precursor proteins for membrane translocation (see above).

Folding mediated by GroEL/ES. The folding intermediate of rhodanese is transferred *in vitro* from the complex with DnaJ-DnaK to GroEL for folding to the native state (Fig. 2, stage 5). Unlike DnaK/DnaJ, GroEL mediates the folding to more compact intermediates in the folding pathway in an ATP-dependent process. The remarkable quarternary structure of GroEL, consisting of two stacked heptameric rings of identical ~60 kDa subunits (Hohn et al., 1979; Hendrix, 1979) enclosing a central cavity, may sign responsible for this capacity. Only one or two folding polypeptides bind per GroEL tetradecamer (Martin et al., 1991b; Viitanen et al., 1991). A recent electron microscopic study revealed that this interaction occurs most likely within the central cavity of the GroEL cylinder (Langer et al., 1992b). It seems possible that hydrophobic motifs are recognized by the chaperone which are buried in the folded state of the substrate protein but are exposed in early folding intermediates. For its full function in folding, GroEL depends on yet another heat shock protein, GroES (Hsp10). This smaller protein, a heptameric ring of 10 kDa subunits (Chandrasekhar et al., 1986), functions as a 'pacemaker' of the folding reaction by regulating the ATP-ase activity of GroEL (Martin.et al., 1991b). A single GroES oligomer binds asymmetrically to one end-surface of the GroEL cylinder (Langer et al., 1992b). Once a folding intermediate associates with GroEL, rounds of ATP driven release and rebinding may take place. Binding of the protein stimulates the ATP-ase of GroEL. In the absence of GroES, ATP-hydrolysis does not result in a permanent release, however, except for certain proteins such as dihydrofolate reductase, which fold spontaneously with high efficiency. As demonstrated with rhodanese, the substrate protein rather undergoes rapid cycles of release and rebinding which can be interrupted by adding an unfolded competitor molecule such as casein, a protein which is only loosely folded and exposes hydrophobic surfaces. Casein competes with rhodanese for re-binding to GroEL and displaces the former when hydrolyzable ATP is present. As a result, the labile folding intermediate of rhodanese will aggregate. Apparently, releasing a substrate protein in the conformation of a compact folding intermediate is not productive for folding to the native state.

Under regulation by GroES, however, GroEL is able to mediate productive folding. The substrate protein is then released from GroEL in a conformation which is able to reach the native state and safely circumvents the danger of aggregation (Fig. 2, stage 5). Regulation by GroES may prevent the complete release of the polypeptide in a single step, resulting in a step-wise release of the protein. Released segments of the protein substrate would thereby be free to fold, whereas other parts of the protein would remain associated with the chaperonin. Alternatively, the substrate protein is completely released but folds within the cavity of GroEL before leaving the protecting chaperonin. Incompletely folded intermediates would be able to re-bind and then undergo another cycle of ATP-dependent release into the GroEL cavity. Such re-binding may overcome unproductive interactions that may occur within the folding polypeptide chain. Thus, folding would proceed in an environment shielded from the cellular surroundings. As expected, the competitor casein does not interrupt the ongoing folding process of rhodanese when GroES is present. The formation of intermediates of rhodanese closer to the native state than the initially bound form could be observed by analyzing the tryptophan fluorescence properties of GroEL bound rhodanese. These conformational intermediates were still in contact with GroEL. There is no evidence that GroEL alters the folding pathway of a protein. Its function seems to be restricted to providing an optimal intracellular microenvironment where aggregation and unproductive side steps are prevented while the folding protein follows its intrinsically determined route to the native state.

## A GroEL-like component in the eukaryotic cytosol

The principle of sequential chaperone action in protein folding could be of general importance. Whether it applies to the folding of newly-synthesized polypeptides in the eukaryotic cytosol is still unclear, however. A structural or functional equivalent of GroEL/Hsp60 has not yet been identified for this compartment. This is surprising in light of the well established role of the chaperonins in protein folding in mitochondria and the bacterial cytosol and given the presence of both Hsp70 and DnaJ homologues in the cytosol of eukaryotic cells (Caplan and Douglas, 1991; Luke et al., 1991; Atencio and Yaffe, 1992). Recently, a chaperonin candidate for the cytosol, t-complex polypeptide-1 (TCP-1), has been proposed (Ellis, 1990; Gupta,

1990; Trent et al., 1991). The TCP-1 protein has a 40% sequence identity with TF55, an archaebacterial stress protein, which forms a double-ring complex containing 8-9 membered rings of 55 kDa subunits and which has some chaperonin properties (Phipps et al., 1991; Trent et al., 1991). TCP-1 is indeed part of a similar ring-complex termed TRiC (for TCP-1 ring complex) which appears to be present in the cytosol of most eukaryotic cells. Interestingly, TRiC, unlike the chaperinins, is hetero-oligomeric and contains 6-8 different subunits with TCP-1 related sequences (Frydman et al., 1992). TRiC functions in the folding and assembly of tubulin and actin, and is alo able to mediate the folding of a monomeric protein such as firefly luciferase. TRiC may indeed have a more general chaperone function in protein folding and its potential to cooperate with Hsp70 in the folding of newly-synthesized polypeptide chains is at present being explored.

## References

Atencio DP, Yaffe MP (1992) MAS5, a yeast homolog of DnaJ involved in mitochondrial protein import. Mol Cell Biol 12:283-291

Beckman RP, Mizzen LA, Welch WJ (1990) Interaction of hsp70 with newly synthesized proteins: implications for protein folding and assembly. Science 248:850-854

Caplan AJ, Douglas MG (1991) Characterization of YDJ1: a yeast homologue of the bacterial dnaJ protein. J Cell Biol 114:609-621

Chandrasekhar GN, Tilly K, Woolford C, Hendrix R, Georgopoulos C (1986) Purification and properties of the groES morphogenetic protein of *Escherichia coli*. J Biol Chem 21:12414-12419

Cheng MY, Hartl FU, Martin J, Pollock RA, Kalousek F, Neupert W, Hallberg EM, Hallberg RL, Horwich AL (1989) Mitochondrial heat-shock protein hsp60 is essential for assembly of proteins imported in yeast mitochondria. Nature 337:620-625

Deshaies RJ, Koch BD, Werner-Washburne M, Craig EA, Schekman R (1988) A subfamily of stress proteins facilitates translocation of secretory and mitochondrial precursor polypeptides. Nature 332:800-805

Ellis RJ (1987) Proteins as molecular chaperones.Nature 328:378-379

Ellis RJ (1990) Molecular chaperones:the plant connection. Science 250:954-959

Flaherty KM, DeLuca-Flaherty, McKay DB (1990) Three-dimensional structure of the ATPase fragment of a 70K heat-shock cognate protein. Nature 346:623-628

Flynn GC, Pohl J, Flocco MT, Rothman JE (1991) Peptide-binding specificity of the molecular chaperone BiP. Nature 353:726-730

Frydman J, Nimmesgern E, Erdjument-Bromage, H., Wall, JS, Tempst, P, Hartl, FU (1992) Function in protein folding of TRiC, a cytosolic ring complex containing TCP-1 and structurally related subunits. EMBO J. 11: 4767-4778

Gething MJ, McCammon K, Sambrook J (1986) Expression of wild type and mutant forms of influenza hemagglutinin: the role of folding in intracellular transport. Cell 46:939-950

Gething MJ, Sambrook J (1992) Protein folding in the cell. Nature 355:33-45

Glick, BS, Beasley, EM, Schatz, G (1992) Protein sorting in mitochgondria. Trends Biol. Sci.203: 453-458

Gupta RS (1990) Sequence and structural homology between a mouse t-complex protein TCP-1 and the chaperonin family of bacterial (GroEL 65kDa heat-shock antigen) and eukaryotic proteins. Biochem International 20:833-841

Hartl FU, Neupert W (1990) Protein sorting to mitochondria: evolutionary conservations of folding and assembly.Science 247:930-938

Hartl FU, Martin J, Neupert W (1992) Protein folding in the cell: the role of molecular chaperones hsp70 and hsp60. Annu Rev Biophys Biomol Struct 21:293-322

Hendrix RW (1979) Purification and properties of groE, a host protein in bacteriophage assembly. J Mol Biol 129:375-392

Hohn T, Hohn B, Engel A, Wurtz M (1979) Isolation and characterization of the host protein groE involved in bacteriophage lambda assembly. J Mol Biol 129:359-373

Kang PJ, Ostermann J, Shilling J, Neupert W, Craig EA, Pfanner N (1990) Requirement of hsp70 in the mitochondrial matrix for translocation and folding of precursor proteins. Nature 348:137-143

Koll H, Guiard B, Rassow J, Ostermann J, Horwich AL, Neupert W, Hartl FU (1992) Antifolding activity of hsp60 couples protein import into the mitochondrial matrix with export to the intermembrane space. Cell 68:1163-1175

Landry SJ, Jordan R, McMacken R, Gierasch LM (1992) Different conformations for the same polypeptide bound to chaperones DnaK and GroEL. Nature 355:455-457

Langer T, Lu C, Echols H, Flanagan J, Hayer MK, Hartl FU (1992a) Successive action of molecular chaperones DnaK, DnaJ and GroEL along the pathway of assisted protein folding.Nature 356:683-689

Langer T, Pfeifer G, Martin, J, Baumeister W, Hartl FU (1992b) Chaperonin-mediated protein folding: GroES binds to one end of the GroEL cylinder, which accommodates the protein substrate within its central cavity. EMBO J. 11: 4757-4766

Luke MM, Sutton A, Arndt KT (1991) Characterization of SIS1, a *Saccharomyces cerevisiae* homologue of bacterial dnaJ proteins. J Cell Biol 114:623-638

Manning-Krieg UC, Scherer PE, Schatz G (1991) Sequential action of mitochondrial chaperones in protein import into the matrix. EMBO J 10:3273-3280

Martin J, Mahlke K, Pfanner N (1991a) Role of an energized inner membrane in mitochondrial protein import. J Biol Chem 266:18051-18057

Martin J, Langer T, Boteva R, Schramel A, Horwich AL, Hartl FU (1991b) Chaperonin-mediated protein folding at the surface of groEL through a 'molten globule'-like intermediate.Nature 352:36-42

Neupert W, Hartl FU, Craig EA, Pfanner N (1990) How do polypeptides cross the mitochondrial membranes? Cell 63:447-450

Ostermann J, Horwich AL, Neupert W, Hartl FU (1989) Protein folding in mitochondria requires complex formation with hsp60 and ATP hydrolysis. Nature 341:125-130

Ostermann J, Voos W, Kang PJ, Craig EA, Neupert W, Pfanner N (1990) Precursor proteins in transit through mitochondrial contact sites interact with hsp70 in the matrix. FEBS Lett 277:281-284

Pfanner N, Hartl FU, Guiard B, Neupert W (1987) Mitochondrial precursor proteins are imported through a hydrophilic membrane environment. Eur J Biochem 169:289-293

Phipps BM, Hoffman A, Stetter KO, Baumeister W (1991) A novel ATPase complex selectively accumulated upon heat shock is a major cellular component of thermophilic archaebacteria. EMBO J 10:1711-1722

Rassow J, Hartl FU, Guiard B, Pfanner N, Neupert W (1990) Polypeptides traverse the mitochondrial envelope in an extended state. FEBS Lett 275:190-194

Rothman JE (1989) Polypeptide chain binding proteins: catalysts of protein folding and related processes in cells.Cell 59:591-601

Scherer PE, Krieg UC, Hwang ST, Vestweber D, Schatz G (1990) A precursor protein partly translocated into yeast mitochondria is bound to a 70 kd mitochondrial stress protein. EMBO J 9:4315-4322

Trent JD, Nimmesgern E, Wall JS, Hartl FU, Horwich AL (1991) A molecular chaperone from a thermophilic archaebacterium is related to the eukaryotic protein t-complex polypeptide-1. Nature 354:490-493

Viitanen PV, Donaldson GK, Lorimer GH, Lubben TH, Gatenby AA (1991) Complex interactions between chaperonin 60 molecular chaperone and dihydrofolate reductase. Biochemistry 30:9716-9723

Wickner S, Hoskins J, McKenney K (1991) Function of DnaJ and DnaK as chaperones in origin-specific DNA binding by RepA. Nature 350:165-167

Zimmerman SB, Trach SO (1991) Estimation of macromolecule concentrations and excluded volume effects for the cytoplasm of *Escherichia coli*. J Mol Biol 222:599-620

# COMPONENTS AND MECHANISMS INVOLVED IN PROTEIN TRANSLOCATION ACROSS THE ENDOPLASMIC RETICULUM MEMBRANE

Tom A. Rapoport
Max-Delbrück-Center for Molecular Medicine
Robert-Rössle-Str. 10
O-1115 Berlin-Buch
FRG

Many proteins are transported across the ER membrane as they are synthesized. These include secretory proteins and proteins of the plasma membrane, of lysosomes, of endosomes, and of all organelles of the secretory pathway. Synthesis of these proteins begins in the cytoplasm, but they are then targeted to the ER membrane by signal sequences, which are characterized by a continuous stretch of at least 6 apolar amino acids, and which are often located at the $NH_2$-terminus of precursor molecules. Recognition of the signal sequence and targeting of the nascent chain generally requires the combined function of the signal recognition particle (SRP) and of its membrane receptor, but alternative targeting pathways exist. The targeting phase is succeeded by the actual translocation process. Proposed mechanisms of translocation have ranged from the idea that the transport of a polypeptide chain occurs directly through the phopholipid bilayer without participation of membrane proteins to models in which polypeptides are transported through a hydrophilic or amphiphilic channel formed from transmembrane proteins (for discussion, see Rapoport (1991)). It now seems likely that a protein-conducting channel does exist. The evidence comes from electrophysiological data and from the identification of membrane proteins as putative channel constituents. Three approaches have contributed to the recent progress - genetic screening for translocation components, identification of membrane proteins adjacent to translocating polypeptides by chemical crosslinking, and reconstitution of translocation components into proteoliposomes after

NATO ASI Series, Vol. H 97
Post-transcriptional Control of Gene Expression
Edited by Orna Resnekov and Alexander von Gabain
© Springer-Verlag Berlin Heidelberg 1996

their solubilization and purification. The present review summarizes our knowledge on the components involved in the actual translocation process. The mechanisms by which polypeptides are targeted to the ER membrane have been discussed elsewhere (Rapoport, 1992).

Polypeptides are transported at specific sites through the ER membrane. The translocation site is probably a rather complex structure, consisting of a number of proteins with different functions. Some of the components of the translocation site may be directly involved in the transport process, others may take part in chemical modifications of a nascent polypeptide or in its folding and assembly. The complexity of the translocation size is indicated by the growing number of components that have been found in it (Table 1).

Table 1. Possible components of the translocation site

| Protein | Function |
|---|---|
| Sec61/SecYp | constituent of a protein-conducting channel |
| TRAM protein | early function in translocation |
| Sec62/63p-complex | early function in translocation |
| SSR-complex | ? |
| signal peptidase complex | signal peptide cleavage |
| oligosaccharyl transferase | Asn-glycosylation |
| mp30 | ? |

Evidence for a protein-conducting channel has been provided recently (Simon and Blobel, 1991). Rough microsomal vesicles fused into planar lipids showed channels of high ion conductivity. The channels increased in number after release of the nascent chains from the ribosomes by puromycin, suggesting that they had been plugged by nascent chains in transit through the ER membrane. They closed if the salt concentration was subsequently raised - conditions known to result in the dissociation of the ribosomes into their subunits. Further evidence for a hydrophilic channel comes from experiments in which the environment of membrane-inserted nascent chains was investigated by measuring the fluorescence life time of incorporated probes (A. E. Johnson, personal communication).

Sec61/SecYp: a core component of the translocation apparatus

Sec61p was discovered in S. cerevisiae in genetic screens for translocation defects (Deshaies and Schekman, 1987). Temperature-sensitive mutations in Sec61p lead to the accumulation of precursor molecules of both secretory and membrane proteins at nonpermissive temperatures.

A mammalian homolog of Sec61p was recently found (Görlich et al., 1992a). 56% of its amino acids are identical with those of the yeast protein. Sec61p can be crosslinked to various translocating secretory proteins in ER membranes from S. cerevisiae (Sanders et al., 1992; Müsch et al., 1992) or canine pancreas (Görlich et al., 1992a). At late stages of the translocation process, when the nascent chains have a sizable lumenal domain, Sec61p is the major protein that becomes crosslinked to the translocating polypeptide. Sec61p is identical with P37, a crosslinking partner of secretory and membrane proteins (High et al., 1991). Thus, Sec61p seems to be closely apposed to polypeptides that are moving through the membrane.

Sec61p has sequence similarity to SecYp of bacteria (Görlich et al., 1992a) (Fig.1). The proteins have identical predicted topologies. Several hydrophilic amino acids within membrane-spanning regions are conserved, suggesting that they are essential for a hydrophilic environment within the membrane.

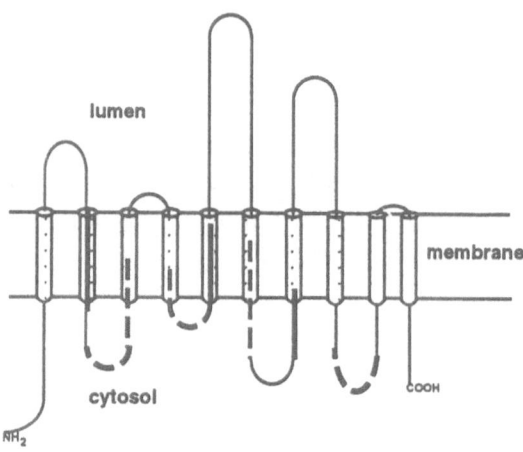

Fig. 1.    Predicted membrane topologies of Sec61p and SecYp. Thick regions indicate sequences with conserved amino acids and high similarity between Sec61p and SecYp from various bacteria, dashed regions indicate sequences of lower but significant similarity.

SecYp is likely to be a major component of the translocation apparatus in bacteria (Schatz and Beckwith, 1990). SecYp together with SecEp may be the only membrane components required for the in vitro translocation of proteins into proteoliposomes (Brundage et al., 1990).

Sec61p has many of the properties expected for a constituent of a protein-conducting channel: (1) It is a neighbor of translocating nascent chains in different organisms; (2) it may form a hydrophilic environment in the membrane; (3) its structure is highly conserved in evolution; and (4) Sec61p from yeast and its prokaryotic counterpart SecYp are essential for translocation in vivo.

Ribosome-association of Sec61p


Mammalian Sec61p is tightly bound to ribosomes after solubilization of rough microsomes in detergent at high salt concentrations (Görlich et al., 1992a). The interaction cannot be exclusively via the nascent chain because the latter can be released from the ribosome by puromycin without causing detachment of Sec61p. However, if the salt concentration is subsequently increased, so that the ribosome dissociates into its two subunits, Sec61p is released. The conditions required to strip rough microsomes from ribosomes are the same as those needed for the dissociation of the isolated Sec61p-ribosome complex. These results are consistent with those from the electrophysiological experiments that provide evidence for a protein-conducting channel. It seems that the closure of the  channel is linked with the dissociation of the ribosomal subunits from Sec61p during termination of translation. The physiological relevance of the interaction between Sec61p and ribosomes is further supported by the fact that the interaction is induced by the targeting of a nascent polypeptide chain to the ER membrane (Görlich et al., 1992a). Ribosomes lacking nascent chains interact more weakly with Sec61p. This reaction probably corresponds to the binding of such ribosomes to microsomes, an assay used in previous studies on ribosome receptors (Borgese et al., 1974).

Because Sec61p is probably a core component of the translocation site, it is likely that its association with the ribosome is caused by a ribosome receptor. The latter may be either Sec61p itself or an associated "adaptor" protein.

A 180-kD protein (Savitz and Meyer, 1990) and a 34-kD protein (Tazawa et al., 1991) have been proposed as ribosome receptors but counterarguments have been raised against both candidates (Nunnari et al., 1991; Görlich et al., 1992a). Of course, more than one ribosome receptor may exist.

The interaction of membrane-bound ribosomes with Sec61p suggests that nascent chains are transferred directly from the channel in the ribosome into a protein-conducting channel in the membrane. This would prevent the premature folding of a polypeptide chain in the cytoplasm into a translocation-incompetent conformation. However, at least in yeast, Sec61p is also involved in the posttranslational translocation of some proteins, such as the secretory protein prepro- α -factor (Sanders et al., 1992; Müsch et al., 1992). Presumably, these proteins maintain a translocation competent state even after their release from the ribosome.


The TRAM protein


Another component of the translocation site is the TRAM protein (Görlich et al., 1992b). Its identification was based on the work of Nicchitta and Blobel (1990), who reconstituted translocation activity in proteoliposomes, and studies that showed that short nascent chains of a secretory protein can be crosslinked to a glycosylated membrane protein (Wiedmann et al., 1987; Krieg et al., 1989). For purification of the crosslinking partner, proteoliposomes with a defined composition of glycoproteins were reconstituted from a cholate extract of canine pancreas microsomes, and tested for the appeerence of a crosslinked product. A single glycoprotein, the "translocating chain associating membrane" (TRAM) protein, turned out to be sufficient to allow crosslinking (Görlich et al., 1992b). The sequence of the TRAM protein, deduced from cloning of the corresponding cDNA, suggests that it spans the membrane eight times and that it has a cytoplasmic tail of about 60 amino acids (Fig.2). Several amino acids in the membrane-spanning regions are hydrophilic or charged. The TRAM protein is about as abundant as ER membrane-bound ribosomes, suggesting that it is present in each translocation site.

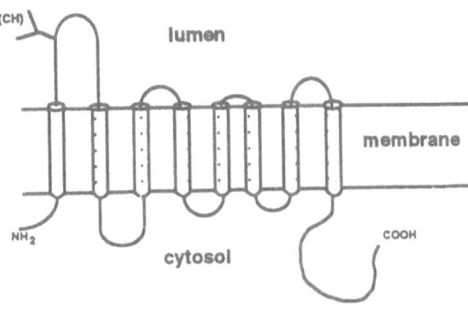

Fig. 2. Predicted membrane topology of the TRAM protein. (CH), Asn-linked carbohydrate chain.

The effect of the TRAM protein on the translocation of secretory proteins was tested in an improved reconstitution system, having an overall transport efficiency approaching that of the original membranes (Görlich et al., 1992b). Proteoliposomes depleted of glycoproteins had a reduced transport activity for prepro-α-factor and pre-ß-lactamase but had an only slightly reduced activity for preprolactin. Addition of the TRAM protein was sufficient to restore translocation to the original levels.

The differential effect of glycoprotein depletion on different translocation substrates is similar to the differential effects of several Sec mutants (Deshaies and Schekman, 1989) and may indicate different requirements for translocation components. However, the depletion of the TRAM protein might not have been complete. The number of translocation sites containing the TRAM protein may have only been reduced and proteins with a high affinity for them, like presumably preprolactin, could have still been translocated.

Short nascent chains of secretory proteins representing early stages of translocation can be crosslinked to both the TRAM protein and Sec61p (Görlich et al., 1992a). The TRAM protein seems to interact with amino acid residues preceding the hydrophobic core of the signal sequence whereas Sec61p seems more in contact with the core and with residues succeding it (High et al., 1993).

With longer chains of secretory proteins crosslinks to the TRAM protein have not been observed. It is possible that these longer chains lack suitably located amino acids for crosslinking, but it seems more likely that the TRAM protein is only adjacent to nascent chains at the beginning of their membrane passage. It seems possible that signal peptide cleavage causes displacement of the TRAM protein.

The Sec62p-Sec63p complex

Sec62p and Sec63p were detected in S. cerevisiae in genetic screens for translocation components (Deshaies and Schekman, 1987). They form a complex that also includes a glycoprotein of 31.5 kD and a non-glycoprotein of 23 kD (Deshaies et al., 1991). There is more Sec63p than Sec62p in yeast cells. Sec63p interacts with BiP (Kar2), a chaperone located in the lumen of the ER. Because temperature-sensitive mutations of BiP result in rapid appearence of translocation defects at nonpermissive temperatures (Vogel et al., 1990), it seems possible that the chaperone directly participates in the transport process, perhaps by pulling the polypeptide chain across the membrane.

The Sec62p-Sec63p complex may function at early stages of the translocation process during which nascent polypeptides crosslink weakly to Sec62p (Müsch et al., 1992). Also, mutations in Sec62p or Sec63p prevent the interaction of translocating chains with Sec61p (Sanders et al., 1992).

Enzymes in the translocation site

Two enzymes are known to be located in the translocation site, the signal peptidase and the oligosaccharyltransferase. They catalyze cotranslational modifications of the polypeptide chain and are unusual enzymes in that they are as abundant as their substrates. A special structural arrangement is probably needed for the enzymes to act on uncompleted nascent chains which may be surrounded by other membrane proteins. It seems likely that the two enzymes are a part of each translocation site as long as it remains assembled.

The signal peptidase consists of five different proteins (Evans et al., 1986). The oligosaccharyltransferase has been purified from dog pancreatic microsomes and consists of three subunits, the two ribophorins (I and II) and a 48-kD polypeptide (Kelleher et al., 1992).

Both the oligosaccharyltransferase and the signal peptidase seem to be dispensable for the actual translocation process because protein translocation occurs in proteoliposomes depleted of all glycoproteins (including both enzymes) except the TRAM protein (Görlich et al., 1992b).

The signal sequence receptor complex

The signal sequence receptor α subunit (SSRα) was purified on the basis of its properties as deduced from its crosslinking to short translocating polypeptide chains (Hartmann et al., 1989). The nascent chains are crosslinked to a protein that is about 35 kD, has a cytoplasmic tail of about 5 kD, is glycosylated, and is abundant (Wiedmann et al., 1987). Both the TRAM protein and SSRα meet these requirements, and the TRAM protein is actually the major crosslinking partner of short nascent chains (Görlich et al., 1992b). However, SSRα can also be crosslinked to various translocating chains, and the proportion of SSRα among the glycoproteins crosslinked to the nascent chain appears to increase as chain length increases (Görlich et al., 1992b). The protein is not a signal sequence receptor and most likely not even directly involved in translocation. Proteoliposomes reconstituted from a detergent extract from which SSR was removed by immuno-affinity chromatography (Migliaccio et al., 1992), and proteoliposomes containing the TRAM protein as the only glycoprotein (Görlich et al., 1992b) both have unreduced translocation activity.

Nevertheless, the SSR is likely to be located in the translocation site. It is associated in part with ribosomes after solubilization of rough microsomes, is segregated to the rough portion of the ER and can be crosslinked to membrane-bound ribosomes (for discussion, see Rapoport, 1991). Antibodies to SSRα and Fab-fragments prepared from the antibodies block the in vitro translocation of several secretory proteins (Hartmann et al., 1989).

SSRα is a constituent of a stoichiometric complex of four membrane proteins (SSR-complex) which are adjacent to each other in intact membranes. The amino acid sequences deduced from cloning of the corresponding cDNAs indicate that the α, ß and δ subunits span the membrane only once. The γ subunit is predicted to span the membrane four times (Fig.3).

The function of the SSR-complex remains unclear. It may have an enzymatic activity or it could be required for the translocation of only a subclass of proteins. Alternatively, it could function as a chaperone, facilitating the folding of membrane proteins.

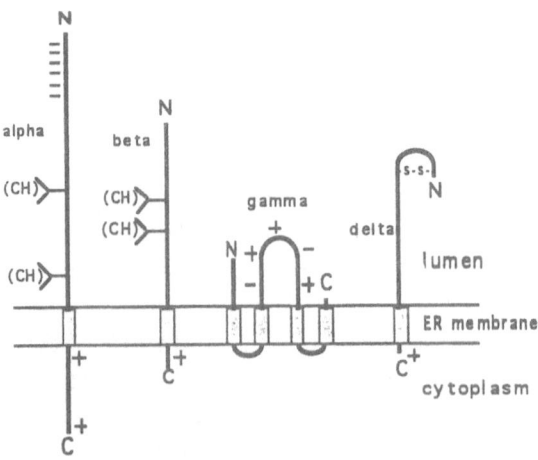

Fig.3. Predicted membrane topologies of the subunits of the SSR-complex. (CH), carbohydrate chains. Plus and minus signs indicate positively and negatively charged polypeptide segments.

Further components implicated in translocation

Other candidates for translocation components include an abundant ER membrane protein of 30 kD (mp30) with affinity for the SRP (Tajima et al., 1986), an unidentified ATP-binding membrane protein (Klappa et al., 1991), and a number of proteins that have been identified as crosslinking partners of translocating nascent chains but are only known by their approximate molecular weights.

Conclusions

It seems now certain that membrane proteins are involved in the translocation process. The translocation site contains not only proteins that are essential for the transport process, but also enzymes that catalyze the modification of nascent polypeptides and probably proteins needed for other functions (see Table 1). A protein-conducting channel is likely to exist.

Sec61p and SecYp seem to be major components of the putative protein-conducting channel. Other membrane proteins, like the TRAM protein or the Sec62p-Sec63p complex may be only transiently required. In general, at least in mammals, the nascent polypeptide seems to be transferred directly from the ribosome into the translocation site through a tight junction between the membrane-bound ribosome and Sec61p.

The mechanisms of protein transport across the ER membrane and across the cytoplasmic membrane in bacteria seem to be basically the same. Both, the discovery of a complex similar to the SRP that seems to function in protein export from bacteria (Phillips and Silhavy, 1992) and the obvious similarity of SecYp and Sec61p, provide a mechanistic correlate to the long known fact that signal sequences are similarly structured and exchangable between different classes of organisms.

The precise mechanism of translocation remains unclear but one may now hope to reconstitute into proteoliposomes the translocation process from purified components.

REFERENCES

Borgese, N., Mok, W., Kreibich, G., and Sabatini, D. (1974). Ribosomal-membrane interaction: in vitro binding of ribosomes to microsomal membranes. J. Mol. Biol. 88, 559-580.

Brundage, L., Hendrick, J. P., Schiebel, E., Driessen, A. J. M., and Wickner, W. (1990). The purified Escherichia coli integral membrane protein SecY/E is sufficient for reconstitution of SecA-dependent precursor protein translocation. Cell 61, 649-657.

Deshaies, R. J., and Schekman, R. (1987). A yeast mutant defective at an early stage in import of secretory protein precursors into the endoplasmic reticulum. J. Cell Biol. 105, 633-645.

Deshaies, R. J., and Schekman, R. (1989). Sec62 encodes a membrane protein required for protein translocation into the yeast endoplasmic reticulum. J. Cell Biol. 109, 2653-2664.

Deshaies, R. J., Sanders, S. L., Feldheim, D. A., and Schekman, R. (1991). Yeast Sec proteins involved in translocation into the endoplasmic reticulum are assembled into a membrane-bound multi-subunit complex. Nature 349, 806-808.

Evans, E., Gilmore, R. and Blobel, G. (1986) Purification of microsomal signal peptidase as a complex. Proc. Natl. Acad. Sci. USA 83: 581-585.

Görlich, D., Prehn, S., Hartmann, E., Kalies, K.-U. and Rapoport, T. A. (1992a) A mammalian homolog of Sec61p and SecYp is associated with ribosomes and nascent polypeptides during translocation. Cell 71, 489-503.

Görlich, D., Hartmann, E., Prehn, S., and Rapoport, T. A. (1992b). A protein of the endoplasmic reticulum involved early in polypeptide translocation. Nature 357, 47-52.

Hartmann, E., Wiedmann, M. and Rapoport, T. A. (1989) A membrane component of the endoplasmic reticulum that may be essential for protein translocation, EMBO-J. 8, 2225-2229.

High, S., Görlich, D., Wiedmann, M., Rapoport, T. A., and Dobberstein, B. (1991). The identification of proteins in the proximity of signal-anchor sequences during their targeting to and insertion into the membrane of the ER. J. Cell Biol. 113, 35-44.

High, S., Martoglio, B., Görlich, D., Andersen, S. S. L., Ashford, A. J., Giner, A., Hartmann, E., Prehn, S., Rapoport, T. A., Dobberstein, B. and Brunner, J. (1993) Site specific photocrosslinking reveals that Sec61p and TRAM contact different regions of a membrane inserted signal sequence., submitted for publication.

Kelleher, D. J., Kreibich, G., and Gilmore, R. (1992). Oligosaccharyltransferase activity is associated with a protein complex composed of ribophorins I and II and a 48 kd protein. Cell 69, 55-65.

Klappa, P., Mayinger, P., Pipkorn, R., Zimmermann, M., and Zimmermann, R. (1991). A microsomal protein is involved in ATP-dependent transport of presecretory proteins into mammalian microsomes. EMBO-J. 10, 2795-2803.

Krieg, U. C., Johnson, A. E. and Walter, P. (1989) Protein translocation across the endoplasmic reticulum membrane: Identification by photocrosslinking of a 39-kD integral membrane glycoprotein as part of a putative translocation tunnel, J. Cell Biol. 109, 2033-2043.

Migliaccio, G., Nicchitta, C. V. and Blobel, G. (1992). The signal sequence receptor, unlike the signal recognition particle receptor, is not essential for protein translocation. J. Cell Biol. 117, 15-25.

Müsch, A., Wiedmann, M. and Rapoport, T. A. (1992). Yeast Sec proteins interact with polypeptides traversing the endoplasmic reticulum membrane. Cell 69, 343-352.

Nicchitta, C. and Blobel, G. (1990). Assembly of translocation competent proteoliposomes from detergent-solubilized rough microsomes. Cell 60, 259-269.

Nunnari, J. M., Zimmerman, D. L., Ogg, S. C., and Walter, P. (1991). Characterization of the rough endoplasmic reticulum ribosome-binding activity. Nature 352, 638-640.

Phillips, G. J. and Silhavy, T. J. (1992) The E. coli ffh gene is necessary for viability and efficient protein export. Nature 359, 744-746.

Rapoport, T. A. (1991). Protein translocation across the endoplasmic reticulum membrane: facts, models, mysteries. FASEB J. 5, 2792-2798.

Rapoport, T. A. (1992) Transport of proteins across the endoplasmic reticulum membrane. Science 258: 931-936.

Sanders, S. L., Whitfield, K. M., Vogel, J. P., Rose, M. D., and Schekman, R. W. (1992). Sec61p and BiP directly facilitate polypeptide translocation into the ER. Cell 69, 353-366.

Savitz, A. J., and Meyer, D. I. (1990) Identification of a ribosome receptor in the rough endoplasmic reticulum. Nature 346, 540-544.

Schatz, O. J. and Beckwith, J. (1990) Genetic analysis of protein export in Escherichia coli. Annu. Rev. Genet. 24: 215-248.

Simon, S. M., and Blobel, G. (1991). A protein-conducting channel in the endoplasmic reticulum. Cell 65, 371-380.

Tajima, S., Lauffer, L., Rath, V. L. and Walter, P. (1986) The signal recognition particle receptor is a complex that contains two distinct polypeptide chains. J. Cell Biol. 103: 1167-1178.

Tazawa, S., Unuma, M., Tondokoro, N., Asano, Y., Ohsumi, T., Ichimura, T., and Sugano, H. (1991). Identification of a membrane protein responsible for ribosome binding in rough microsomal membranes. J. Biochem. 109, 89-98.

Vogel, J. P., Misra, L. M., and Rose, M. D. (1990) Loss of BiP/GRP78 function blocks translocation of secretory proteins in yeast. J. Cell Biol. 110: 1885-1895.

Wiedmann, M., Kurzchalia, T. V., Hartmann, E., and Rapoport, T. A. (1987). A signal sequence receptor in the endoplasmic reticulum membrane. Nature 328, 830-833.

# EXPORT OF PROTEINS FROM BACTERIA AND TRANSLOCATION OF PROTEINS ACROSS THE ENDOPLASMIC RETICULUM OF EUKARYOTIC CELLS; A COMPARISON

Simon Hardy
Department of Biology
University of York
York YO1 5DD
U.K.

## Introduction

All proteins of bacteria and the vast majority of proteins of eukaryotes are synthesised in the cytoplasm. To perform their function however, many proteins need to be in other locations in the cell, and this almost always involves crossing at least one membrane. I know of no cases where protein localisation is used as a controlling step for gene expression, i.e. where the protein is made and the control of its function is exerted by the degree of its translocation to its final destination. On the contrary, our current view is that virtually all molecules of a protein arrive at their predestined location, any shortfall being an aberration rather than the result of a control mechanism. Therefore the topic of this article is only peripherally relevant to the subject of the course. Nevertheless protein localisation can be viewed as the final step of gene expression without which cells would be in a very sorry state indeed.

It is certain that as soon as proteins and membrane enclosed entities, primitive cells, had appeared in the early environment, the property of being able to communicate with and locally modify that environment would have conferred a big advantage. Thus one would expect that translocation of proteins across membranes would be an extremely fundamental process that would have evolved very early. Since both membranes and proteins have remained essentially the same in bacteria and eukaryotes, it seems likely that membrane translocation of proteins in the two kingdoms would have retained sufficient similarity to be recognisably the same process.

In general terms this expectation is met. Proteins destined for export from bacteria and those to be transferred

NATO ASI Series, Vol. H 97
Post-transcriptional Control of Gene Expression
Edited by Orna Resnekov and Alexander von Gabain
© Springer-Verlag Berlin Heidelberg 1996

from the cytoplasm of eukaryotic cells across the membrane of the endoplasmic reticulum to other compartments, are made with amino terminal leader or signal sequences that are removed during translocation of the proteins across the respective membranes. These extra sequences are essentially identical in bacteria and eukaryotes (von Heijne 1985) to the extent that proteins exported by bacteria can be secreted by eukaryotes although the converse is less often true. Thus one would expect that understanding derived from studies of either system might be directly applicable to the other. For some time, even though each system provided clues for investigation of the other, this hope appeared overoptimistic. However, the recent elucidation of homologies between the bacterial and eukaryotic apparatus for targeting and membrane translocation of proteins has regenerated its promise (Poritz et al. 1990, Phillips and Silhavy, Gorlich et al, 1992), while at the same time increasing the complexity of the models.

It should be emphasised that these models are limited by the experimental systems that give rise to them. For example, the cell free system from dog pancreas that has largely yielded the results on which the signal hypothesis is based, is dependent on studies with rather few proteins. Furthermore if some important component of the system _in vivo_ were labile and thus inactivated during the preparation of the cell free system, its contribution to targeting or translocation would not be detected unless of course it was indispensable _in vitro_. Similarly the genetic studies in _E. coli_ that identified the export apparatus were based on initial selections for mutants defective in the export of only a very few different proteins. If other proteins used a different apparatus, it would not be identified.

With these strictures in mind the signal hypothesis, by far the most influential model and the one found in all text books, will be described and its application to all eukaryotic systems evaluated. There then follows an account of bacterial protein export and finally an evaluation of the discrepancies to see if the two models can be reconciled.

## The signal hypothesis

The major emphasis of the signal hypothesis is on the targeting step of secretion rather than on translocation of the protein across the endoplasmic reticulum membrane. I shall however briefly outline the latter step for purposes of comparison with the bacterial process. For a fuller account see Rapoport 1991, 1992.

Signal sequences in secretory protein precursors are almost always amino terminal extensions that are removed during translocation of the endoplasmic reticulum membrane. They have no defined sequence, but all contain an uninterrupted hydrophobic stretch of at least six and less than fifteen residues which is essential for their function. When the signal sequence emerges from the ribosome it is recognised by and triggers the binding of signal recognition particle (SRP), a complex of six proteins and 7S RNA, to the ribosome and nascent chain. The SRP in its bound state interacts with an integral membrane protein, the docking protein, which releases it from the ribosome, thus positioning the latter on the membrane so that the nascent protein can be translocated cotranslationally. Hydrolysis of GTP is required for release of SRP from the membrane. Electrophysiological evidence that ion conducting pores in the membrane are created by release of nascent chains from membrane-bound ribosomes suggests that the polypeptides are translocated through aqueous pores formed by membrane proteins (Simon & Blobel, 1991). It has not yet been shown that the pores exist before the release of the nascent chains, nor that polypeptides can go through them. However cross-linking studies of translocating chains indicate their close proximity to membrane proteins. During translocation the signal sequence is removed so that it is the mature protein that emerges into the lumen of the endoplasmic reticulum.

The essence of the signal hypothesis is that the targeting signal is directly recognised and that the recognition results in delivery to the correct membrane so that the protein can be translocated cotranslationally. One would therefore predict that an equivalent structure to SRP would be found in all organisms. Although it initially proved elusive in spite of intensive genetic studies, a yeast equivalent of SRP was eventually found by searching for sequence homologies. Gene disruption experiments surprisingly showed that neither the yeast SRP RNA nor the yeast protein homologue of the mammalian polypeptide thought to recognise and bind the signal was essential for viability (Hann & Walter, 1991). When one or both genes were inactivated the yeast grew very slowly but survived. A survey of the fate of a number of proteins that in wild type yeast are rapidly transported through the endoplasmic reticulum membrane and then glycosylated and/or cleaved showed a spectrum of effects from a drastic slowing of processing to wild type behaviour. In addition a retardation of mitochondrial import was observed for the single mitochondrial protein examined, a surprising finding, since studies _in vitro_ of protein import into yeast mitochondria have never shown a requirement for SRP. These data suggest, first that SRP may be only one of a number of targeting systems in yeast such that each individual protein can use more than one but does so preferentially, and second that SRP may not be as specific for signal recognition as had previously been thought but may be able to bind to other kinds of sequences such as mitochondrial import sequences. In other words SRP may be more of a general chaperone than a specific one. (Of course it is possible, even likely, that the effect on mitochondrial protein import is indirect). In addition, the possibility is raised by analogy, that SRP is only one of a number of entities that can facilitate targeting of secretory proteins to the endoplasmic reticulum of mammals. Indeed posttranslational, SRP-dependent, translocation of some small proteins has been observed in the mammalian cell-free system. A similar finding for some rather larger proteins has been made in yeast cell-free systems. (Zimmerman and Meyer, 1986)

Several years ago homologues of two components of SRP were also found in E. coli. (Poritz et al. 1990, Ribes et al. 1990) This was unexpected since neither genetic investigations nor biochemical studies _in vitro_ which together had led to identification of many proteins involved in bacterial protein export, had implicated either component. Both components, the 4.5S RNA and Ffh (Fifty four homologue - the 54 kDA protein that interacts with the signal) were shown to be essential for viability. However no significant effect on protein export could be shown during growth in the absence of 4.5S RNA synthesis.  Finally a similar experiment with Ffh has shown that if a culture is grown in the absence of Ffh synthesis, its growth rate slows after about five generations and the export of the several proteins that were surveyed is retarded to various extents (Phillips and Silhavy, 1992).  Furthermore Ffh was shown to be present in a complex with the 4.5S RNA and the complex was shown to bind to eukaryotic ribosomes carrying a nascent chain with a signal sequence but not if the signal sequence was drastically mutated (Luirink et al. 1992).  Since none of the studies of cell free systems and none of the genetic studies have implicated either 4.5S RNA or Ffh in protein export it is probable that here also, the 'bacterial SRP' is one of a number of cytoplasmic chaperones that can deliver proteins to the export apparatus in the membrane.

## Bacterial protein export

In some ways there are advantages to studying translocation of proteins across membranes in E.coli.  Not only is there a range of powerful and versatile genetic techniques available, but there is also the possibility of carrying out rapid biochemical experiments in vivo so that short-lived states of molecules may be observed.  Bacterial cell free systems for in vitro work have lagged somewhat behind the dog pancreas system but are now better defined in that they can be reconstituted using liposomes and purified proteins.  A current model of protein export in E.coli follows.  For reviews see Randall et al. 1987, Schatz and Beckwith 1990, and Wickner et al. 1991.

Periplasmic and outer membrane proteins of E.coli are synthesised with leader sequences that are essentially identical to signal sequences.  The hydrophobic part of a leader sequence is particularly important since mutations that insert a charge into it normally prevent export.  Either during translation or after it is completed the newly made protein associates with a chaperone, the best understood and most clearly implicated being Sec B, which prevents it from folding into a structure that is incompatible with passage across the membrane.  (Once folded into their native conformations, proteins cannot be translocated.)  There is no evidence that the chaperone recognises the leader sequence, except in the case of the SRP homologue. The chaperone delivers the protein, either nascent or completed, to the export apparatus on the membrane where translocation occurs either co- or post-translationally. There appears to be no significant difference between these two modes of translocation and both are used for nearly all proteins (Josefsson and Randall, 1983). However even when cotranslational, translocation occurs late in synthesis; it does not begin till about 80% of the molecule has been completed (Randall, 1983).

As in eukaryotes, translocation remains mysterious. It involves a peripheral membrane protein Sec A, that hydrolyses ATP in the presence of the translocatory protein, membranes and the integral membrane proteins Sec Y, Sec E and Band I that together make up the translocation apparatus. Sec A probably recognises the leader sequence and initiates translocation. Given enough ATP it can also complete translocation in vitro, but in vivo protein motive force is necessary. In vitro, energising the membrane accelerates translocation drastically and much less ATP is hydrolysed. It is also possible to show that if the membrane polarity is reversed after translocation has started, the protein reverses its direction of movement (Driessen 1992). During translocation the leader sequence is removed by leader peptidase. At least two other proteins Sec D and Sec F have been implicated in protein export by genetic studies but are not required in vitro, a discrepancy which highlights the dangers of relying on a single method of investigation.

SecY is an interesting protein. Implicated very early by genetic studies, it probably spans the membrane ten times and is the leading contender for the translocase or pore forming protein, should such a pore exist. Homologues of SecY have been shown by genetic and cross-linking studies to be involved in yeast protein translocation (SEC61 protein) and also recently in mammalian protein translocation where the homologue can be cross linked to the secretory protein during its traversal of the microsomal membranes in vitro. Furthermore mammalian Sec61p binds tightly to mammalian ribosomes and seems to be responsible for their association with the membrane in the rough endoplasmic reticulum. SecY homologues have also been found in

chloroplasts and in Archaebacteria. The impressive conservation
of Sec Y probably indicates that the mechanism of translocation
is similar in all species (Gorlich et al. 1992).

## Comparisons, questions and conclusion

There appears to be only one fundamental point of
difference between membrane translocation in eukaryotes and
prokaryotes: the source of energy. In bacteria it seems clear
that both nucleoside triphosphates and proton motive force (in
either form, $\Delta$pH or $\Delta\psi$ ) are needed whereas in eukaryotes,
nucleoside triphosphates alone seem sufficient. Uncouplers have
no effect on the eukaryotic cell free systems and no membrane
potential has been detected in the vesicles used. Since the
membrane across which the proteins are transferred in bacteria
is necessarily energised while that in eukaryotes is probably
not, it may be that the eukaryotic system has been forced to
develop its use of nucleoside triphosphates. We may note that
this may not have required too drastic a change, as in vitro the
bacterial system for translocation can work in the absence of
proton motive force, albeit expensively and rather slowly
(Driessen, 1992). A less plausible explanation of the energy
source difference is that it is an artefact, and eukaryotes do
use proton motive force for protein translocation. Such an
explanation, though barely tenable, would be based on the
observation that proton pumps have been found in the endoplasmic
reticulum (Rees-Jones & Al-Awqati, 1984), and that membrane
translocation in eukaryotes has largely been investigated in
cell free systems in which, as in their early bacterial
counterparts, a requirement for energised membranes may be
difficult to demonstrate.

Other differences may be more apparent than real. The
cotranslational/posttranslational dichotomy probably is not
absolute and may well be due to the bias introduced by the use
of particular methodology as discussed above. The apparent lack
of homology between the equivalent components of the targeting
and translocation apparati has begun to look less important with
the finding of bacterial SRP and the SecY/ Sec61p relationship.
As our ignorance diminishes we may expect to find further
congruence in function if not in sequence between the eukaryotic
and prokaryotic components.

Fundamental questions about the mechanism of protein translocation remain:

1) What is the role of the leader/signal sequence at the membrane? In bacteria proteins with mutated leaders are initially transported to the membrane but are not exported. Therefore the leader has some important function after delivery to the membrane. Its hydrophobicity may allow it to partition into the bilayer.

2) Does the protein go through an aqueous pore or through the hydrophobic bilayer? As mentioned above conductivity measurements on membranes favour the former but the data are not conclusive.

3) How is the energy expenditure coupled to translocation of the protein? Since in bacteria both the hydrogen ion potential and the electrical potential are equally effective, the coupling may involve proton antiport or hydroxyl symport. The coupling to nucleoside triphosphate hydrolysis remains obscure.

4) Does the bulk of the protein apart from the leader have a function? Many non translocated proteins when attached to a leader or fused to an exported protein cannot be translocated. Sometimes hydrophobic 'stop transfer' sequences can be identified but often the reason is unknown. These observations may simply mean that translocated proteins do not contain certain difficult sequences, or have evolved to interact more stably with specific chaperones that prevent them from folding into a native or non-translocatable structure, but other more active roles in translocation for the body of the protein are also possible.

5) A related question is: what is the structure of the protein as it translocates? Is it in an extended structure or does it contain secondary structure elements? We know that native proteins cannot be translocated. There is evidence that the precursor to the E.coli protein OmpA when delivered to the membrane contains most of the secondary structure of native OmpA (Lecker et al. 1990).

6) Is there a discrete step for release of a soluble translocated protein from the membrane? There is evidence that some E.coli periplasmic proteins after removal of the leader

sequence remain associated with the cytoplasmic membrane for a short time before being released to the periplasm. SecD and SecF, two integral membrane proteins in search of a specific role in export are known to have large periplasmic domains (Gardel et al. 1990). Their function may be in releasing the translocated protein from the membrane.

The final conclusion then, is that the mechanism of translocation of proteins across membranes is likely to be fundamentally the same in all organisms but that we still have much to learn before we can be sure that our surmise is correct.

## References

Driessen AJM (1992) Bacterial protein translocation: kinetic and thermodynamic role of ATP and the protonmotive force. Trends Biochem Sci 17:219-223

Gardel C, Johnson K, Jacq A and Beckwith J (1990) The SecD locus of E.coli codes for two membrane proteins required for protein export. EMBO J 9:3209-3216

Gorlich D, Prehn S, Hartmann E, Kalies K-U and Rapoport TA (1992) A mammalian homolog of Sec61p and SecYp is associated with ribosomes and nascent polypeptides during translocation. Cell 71:489-503

Hann BC and Walter P (1991) The signal recognition particle in S. cerevisiae. Cell 67:131-144

Joseffson L-G and Randall LL (1981) Different exported proteins in E.coli show differences in the temporal mode of processing in vivo. Cell 25:151-157

Lecker S, Driessen AJM and Wickner W (1990) ProOmpa contains secondary and tertiary structure prior to translocation and is shielded from aggregation by association with SecB protein. EMBO J 9:2309-2314

Luirink J, High S, Wood H, Giner A, Tollervey D and Dobberstein B (1992) Signal-sequence recognition by an Escherichia coli ribonucleoprotein complex. Nature 359:741-743

Phillips GJ and Silhavy TJ (1992) The E.coli ffh gene is necessary for viability and efficient protein export. Nature 359: 744-746

Poritz MA, Bernstein HD, Strub K, Zopf D, Wilhelm H and Walter P (1990) An E.coli ribonucleoprotein containing 4.5S RNA resembles mammalian signal recognition particle. Science 250: 1111-1117

Randall LL (1983) Translocation of domains of nascent periplasmic proteins across the cytoplasmic membrane is independent of elongation. Cell 33:231-240

Randall L, Hardy S and Thom J (1987) Export of protein: a biochemical view. Annu Rev Microbiol 41:507-541

Rapoport TA (1991) Protein transport across the endoplasmic reticulum membrane: facts, models, mysteries. FASEB J 5:2792-2798

Rapoport TA (1992) Transport of proteins across the endoplasmic reticulum membrane. Science 258:931-936

Rees-Jones R and Al-Awqati Q (1984) Proton-translocating adenosine triphosphatase in rough and smooth microsomes from rat liver.  Biochemistry 23:2236-2240

Ribes V, Romisch K, Giner A, Dobberstein B and Tollervey D (1990) E.coli 4.5S RNA is part of a ribonucleoprotein particle that has properties related to signal recognition particle.  Cell 63:591-600

Schatz P and Beckwith J (1990) Genetic analysis of protein export in E.coli.  Annu Rev Genet 24:215-248

Simon SM and Blobel G (1991) A protein-conducting channel in the endoplasmic reticulum.  Cell 65:371-380

von Heijne G (1985) Signal sequences: the limits of variation. J molec Biol 184:99-105

Wickner W, Driessen A and Hartl F-U (1991) The enzymology of protein translocation across the E.coli plasma membrane. Annu Rev Biochem 60:101-124

Zimmerman R and Meyer DI (1986) 1986: a year of new insights into how proteins cross membranes.  Trends Biochem Sci 11:512-515

# ANTISENSE RNA

Kurt Nordström, Stanley N. Cohen[a]), and Robert W. Simons[b])
Department of Microbiology,
Biomedical Center,
Uppsala University,
Box 581, S-751 23
Uppsala
Sweden

## INTRODUCTION

Antisense RNAs are small, diffucible transcripts that are complementary to specific target RNAs. When the antisense and target RNAs pair to one another, target RNA function is altered or inhibited. Natural antisense RNA systems are abundant in prokaryotes and their existence in eukaryotes has been claimed. Artificially engineered antisense RNAs are often used to control gene expression in eukaryotes, especially where classical genetic approaches are difficult or unavailable. Natural and artificial antisense RNA control has been reviewed extensively (Simons and Kleckner, 1988; Simons, 1991; Mol and van der Krol, 1991; Eguchi et al., 1991; Erickson and Izant, 1992; Simons, 1993),

Antisense RNAs control a number of different biological processes in prokaryotic cells and their accessory elements (see Table I). These processes include DNA replication, transposition and conjugal transfer - where quantitative control is achieved - and bacteriophage development and postsegregational killing - where a biological switch is involved. In nearly all of these systems the antisense and target RNAs are transcribed from opposing promoters on the same stretch of DNA. Thus, the RNAs are completely complementary. In a few cases (e. g. ompF and dicF), the antisense and target RNAs are transcribed from

----

[a]) Department of Genetics, Stanford University School of Medicine. Stanford, CA 94305-5120
[b]) Department of Microbiology and Molecular Genetics, 5304 Life Sciences Building, University of California Los Angeles, CA 90024-1489.

NATO ASI Series, Vol. H 97
Post-transcriptional Control of Gene Expression
Edited by Orna Resnekov and Alexander von Gabain
© Springer-Verlag Berlin Heidelberg 1996

different DNA segments and their complementarity is interrupted. In all carefully examined cases, the antisense RNAs are highly structured, being comprised of one or more double-stranded stems each topped by single-stranded loops. These structural motifs are important features of efficient and specific antisense RNA control. In the few cases examined in detail, target RNA structure is also important.

**Table I.** Some cases of natural antisense RNA control in prokaryotes[a]

| Biological system | Activity controlled | Mechanism |
| --- | --- | --- |
| ColE1 plasmids | Replication | Indirect inhibition of preprimer structure and maturation |
| IS10 | Transposition | Direct inhibiton of ribosome binding |
| R1 plasmid | Replication | Translational attenuation |
| pT181 | Replication | Transcriptional attenuation |
| Phage λ | cII expression | Transcript stability |
| Phage P22 | Antirepression | Inhibition of translation |
| E. coli micF/ompF | Osmoregulation | Inhibition of translation |
| R1 and F plasmids | Postsegregational killing | Inhibition of translation and/or mRNA stability |
| E. coli dicF/ftsZ | Cell division | Inhibition of translation |
| R1162 plasmids | Replication | Inhibition of translation? |
| pSL1 plasmid | Replication | Inhibitionof translation? |
| IncF plasmids | Conjugation | Inhibition of translation? |
| Plasmid R6K | Origin activity | Uncertain |
| Phage λ | Q expression | Uncertain |
| E. coli sulA/isf | SOS recovery? | Uncertain |

a) For key references, see Eguchi et al., 1991; Simons 1991 and 1993.

Antisense RNA control can occur at any of several posttranscriptional levels, either "directly" or "indirectly", depending on whether the site of duplex formation is the same as the actual site of control, or is located some distance away. These various mechanisms are well illustrated by the better-characterized systems. The plasmid pT181 antisense RNA controls replication by inhibiting expression of the essential *repC* function in a mechanism reminiscent of transcriptional attenuation of the *Escherichia coli* tryptophan operon. During the synthesis of the RepC mRNA, the antisense RNA binds to and sequesters an upstream mRNA sequence, promoting the formation a Rho-independent transcriptional terminator and, thereby, premature termination of mRNA synthesis (Novick *et al.*, 1989). This is an example of "indirect" inhibition.

Antisense RNA can also facilitate mRNA decay. The bacteriophage λ *c*II OOP RNA, whose significance has long been disputed, pairs to the 3' end of the λ *c*II mRNA, leading to its destabilization in a reaction that depends on the *E. coli* double-stranded-RNA-specific endonuclease, RNase III (Krinke and Wulff, 1987). This is an example of "direct" antisense RNA control.

Control of plasmid ColE1 replication (described in greater detail below) is indirect. The antisense RNA inhibits processing of the primer by altering preprimer structure in a region far away from the duplex. In IS*10* and plasmid R1 (both also described below), antisense RNAs inhibit translation. In IS*10*, transposase translation is inhibited directly, by duplex formation at the ribosome-binding site. In plasmid R1, translation of the essential *repA* function is inhibited indirectly, by duplex formation at a region adjacent to the ribosome-binding site of an upstream gene, whose translation is coupled to that of the downstream *repA* gene. In these and all other known cases, antisense RNA control is negative. However, this need not always be the case. For example, an antisense RNA might prevent the formation of, say, a transcriptional terminator, and thereby activate gene expression.

Below, we discuss several aspects of natural antisense RNA control in greater detail. Three systems will be treated in some detail: control of replication of plasmids ColE1 and R1 and control of transposition of IS10. The general properties of the systems and the kinetics of the interaction between antisense and target RNAs will be discussed. This is followed by a brief accounting of possible cases of antisense RNA control in eukaryotes, and we close with a discussion of how the natural cases provide some lessons for the design of efficient and specific artificial antisense RNA control.

# REPLICATION CONTROL IN ColE1-TYPE PLASMIDS BY ANTISENSE RNA

The first known instance of biological control by antisense RNA and also the most extensively characterized is the RNAI/RNAII system that regulates replication of ColE1 and related *E. coli* plasmids such as pBR322, pACYC184, and pUC replicons. The extensive investigations of Tomizawa and his associates (Eguchi *et al.*, 1991), and also work from the laboratories of Polisky (Polisky, 1988) and Cesareni (Cesareni and Banner, 1985) have established the characteristics of replication control in these ColE1-type plasmids. The information obtained has in turn provided a conceptual framework for the study of other instances of genetic regulation mediated by antisense RNA.

ColE1 replication requires a plasmid-encoded 555 nucleotide (nt) RNA molecule, RNAII, which serves as a primer for the synthesis of plasmid DNA (Figure 1). To become an active primer, RNAII must first form a hybrid with its template DNA by hydrogen bonding, and then be cut by the enzyme RNaseH, which has the ability to digest the RNA component of DNA/RNA hybrids. Cleavage of DNA-bound RNAII at a specific site produces a free 3' OH end; deoxyribonucleotides are added to this terminus by DNA polymerase I, initiating a round of unidirectional plasmid DNA synthesis at the cleaved site (*i. e.* the replication origin). A second RNA species, RNAI is transcribed in an orientation opposite to that of the RNAII transcript from a 108 *nt* segment of the same sequence of plasmid DNA that encodes RNAII. Interaction of this antisense RNA with RNAII causes conformational changes in the primer that inhibit its binding to template DNA and consequently prevent the initiation of DNA synthesis (Fig. 2B and C).

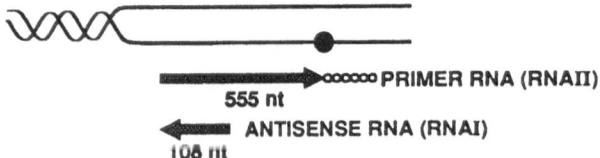

Figure 1. Interaction of antisense RNAI with RNAII to control plasmid DNA replication. • indicates the replication origin for leading strand synthesis of plasmid DNA. The small open circles indicate newly synthesized leading strand DNA.

Multiple factors can affect the regulation of plasmid DNA synthesis by RNAI. Structural probing of RNAI and RNAII using enzymes that cleave specifically in single- and double-stranded regions has shown that both of these RNA species have extensive secondary structure (Figure 2A and B). Mutations that affect the ability of sequences within RNAI or RNAII to pair intramolecularly and form stem-loop structures can alter the conformation of these molecules and consequently modify RNAI/RNAII interactions (Tomizawa, 1984; Tomizawa and Som, 1984). Additionally, mutations affecting RNAII outside of the region of complementarity with RNAI can affect the RNAI/RNAII interaction, probably also by altering the conformation of the primer (Fitzwater *et al.*, 1988; Lin-Chao *et al.*, 1992). Importantly, a small (63 amino acids) ColE1-encoded protein initially called Rop (Repressor of primer) but subsequenty renamed Rom (RNA-one modulator) has been found to stabilize the formation of intermolecular complexes between the stem-loop regions of RNAI and RNAII, leading to a decreased rate of plasmid replication and a lower plasmid copy number.

Plasmid incompatibility, which reflects the inability of replicons to coexist stably in the absence of selection in a clonal cell line, is a measure of the relatedness of plasmids. For ColE1-type plasmids, incompatibility reflects the ability of RNAI from plasmid A to interact with, and inhibit priming by, the RNAII of plasmid B. It has been shown that this is determined largely by sequences within stem-loop regions, which initiate complex formation by weak interaction between the loops, a process that Tomizawa has called "kissing". This interaction leads to increasingly tight binding between RNAI and RNAII, altering the conformation of the primer and consequently preventing duplex formation with the DNA template and cleavage by RNaseH. The loop segments of the stem loops of ColE1 RNAI can interact with the loops of homologous RNAII, but not with the loops of plasmids related to, but compatible with, ColE1. Recent evidence indicates that incompatibility of ColE1-type plasmids is not determined solely by interactions between the stem-loop structures, as has been believed (Lin-Chao and Cohen, 1991); sequence changes in the single-stranded region near the 5' end of RNAI can also alter the incompatibility properties of this group of plasmids, consistent with earlier evidence that this region has an important role in the formation of RNAI/RNAII complexes.

The effectiveness of an antisence RNA as a mechanism for gene regulation depends not only on factors that affect binding of the antisense molecules with their target, but also on factors that affect the relative cellular concentrations of antisense and target molecules. In recent years, it has become evident that the rate of decay of antisense RNA is a key element of antisense RNA control systems. In addition to its major role in elucidating the parameters that affect antisense RNA binding to the target, RNAI has several important features that have made it a highly useful tool for studying ribonucleolytic decay. Because RNAI functions as a regulator of plasmid DNA replication, its biological activity can be readily assessed and mutations that affect its decay *in vivo* can be selected. Since RNAI is not translated, possible confounding effects of translation on the rate of decay are absent.

Antisense RNAI is cleaved by the endonuclease RNaseE at a site five nt from its 5' end (Tomcsanyi and Apirion, 1985), converting relatively stable pRNAI into a rapidly-decaying product, pRNAI-5 (Lin-Chao and Cohen, 1991)(Figure 3). Nucleotide substitutions in the sequence cleaved by RNaseE can affect the rate of cleavage and change the copy number of the plasmid. Temperature-sensitive mutations in the *rne/ams* gene, which encodes or controls the synthesis of RNaseE, yield an increase in the copy number of RNAI-regulated replicons at non-permissive temperature (Table II). Alteration of the pBR322 sequence in the region of the site of cleavage affects the rate of cleavage and consequently alters repression of primer formation by the antisense RNA, the rate of plasmid DNA replication, and the cellular concentration of plasmid DNA.

A triphosphate-terminated mutant homolog (pppRNAI-5) of RNAI-5 is much more stable that pRNAI-5; as the sequences of pppRNAI-5 and pRNAI-5 are identical, it has been suggested that differentially rapid decay observed for RNAI-5 is mediated by a 5' to 3' exonuclease that does not act on the mutationally produced pppRNAI-5, which lacks an RNaseE cleavage site (Lin-Chao and Cohen, 1991). It presently is unclear whether the putative exonucleolytic activity that degrades pRNAI-5 is an ancillary function of the same RNaseE protein that generates this substrate, or is instead due to a separate enzyme.

The RNAI/RNAII system has served as an important model for elucidating the factors that affect the interaction between an antisense RNA species and its target. However, investigations of antisense RNA

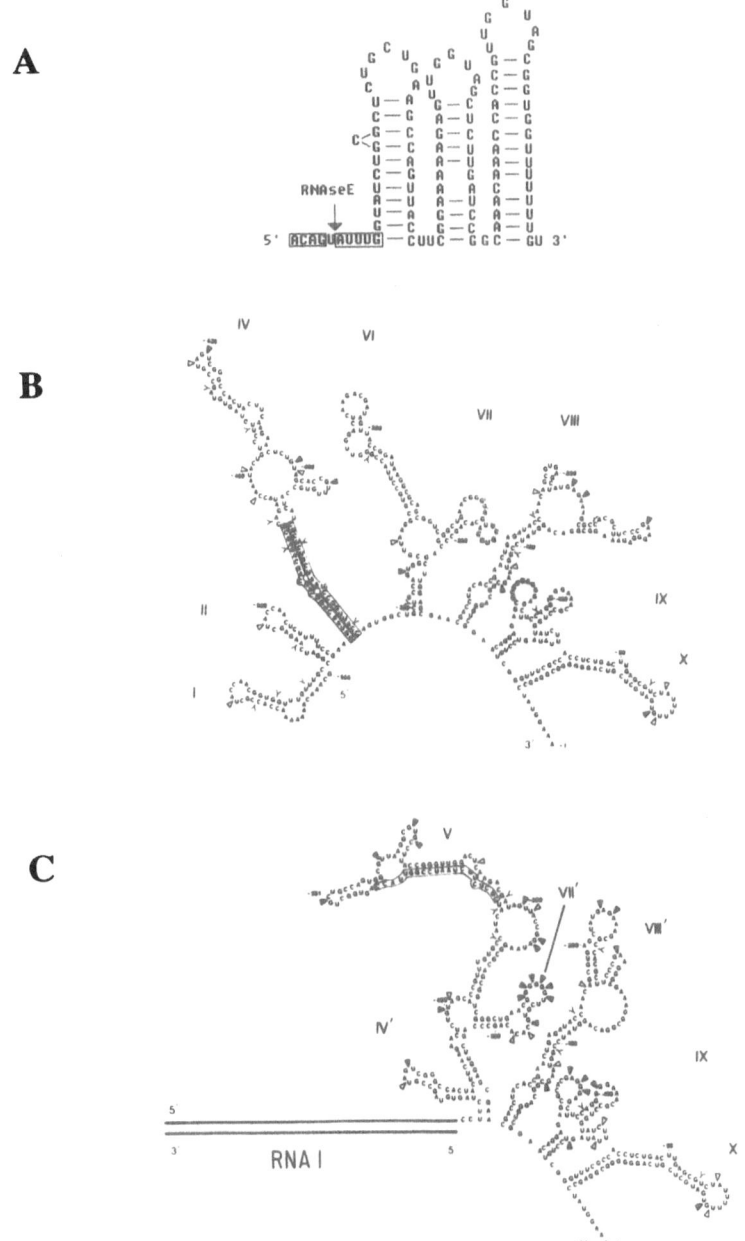

**Figure 2.** A: Primary and secondary structure of RNAI of plasmid pBR322. B:Structure of RNAII in the absence of RNAI. C. Structure of RNAII in the presence of RNAI (from Eguchi *et al.*, 1991).

238

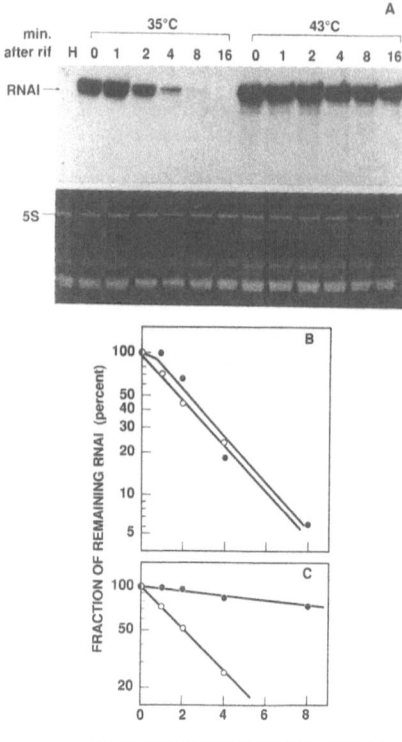

**Figure 3**. Decay of pBR322 RNAI in wild-type *E. coli vs* a *ts* mutant defective in RNaseE activity, as shown by Northern Blot analysis (Lin-Chao and Cohen, 1991). (A) Autoradiograph of electrophoretically separated RNA from *rne(ts)* strain grown at the temperatures shown and hybridized with a ribonucleotide probe containing a sequence complementary to RNAI of pBR322. (B) Plot of the total area under the peak obtained by densitometric scanning of the autoradiographs of RNAI extracted from an *E. coli rne*+ strain (open circles) or *rne* strain (closed circles) grown exponentially at 35 °C. (C) Plot of the peak area obtained by densitometric scanning of the autoradiographs of RNAI extracted from an *rne*+ (open circles) or *rne(ts)* (closed circles) strain after temperature upshift to 43°C for 30 min.

inhibition as a mechanism of genetic regulation *in vivo* require consideration also of the stability of both of these components of an antisense RNA system. Antisense RNA control of the initiation of replication of ColE1-type plasmids *in vivo* is governed significantly by the rate of degradation of RNAI. While the product of initial cleavage of RNAI by the endonuclease RNaseE can interact with the replication primer, RNAII, cleavage of RNAI converts the slowly degraded intact molecule into a rapidly decaying product, relieving antisense RNA repression of plasmid DNA replication. The rate of cleavage of RNAI is a major factor that determines the effective concentration of RNAI; together with the ColE1-encoded Rom protein, which affects the binding of RNAI to RNAII, and the conformations of RNAI and RNAII, the rate of decay of RNAI has a key role in regulating the extent of antisense RNA inhibition of plasmid DNA replication.

**Table II.** Effect of RNaseE inactivation on plasmid DNA concentration in an *rne(ts)* host (Lin-Chao and Cohen, 1991).

| Time (hr) after shift to 43 °C | Plasmids per $UD_{600}$ Unit[a] | | | |
|---|---|---|---|---|
| | pBR322 | ColE1 | pZC20 | pZCLC20 |
| 0 | 1.0 | 1.0 | 1.0 | 1.0 |
| 0.5 | 0.7 | 0.73 | 1.1 | 0.85 |
| 1.0 | 0.54 | 0.46 | 1.1 | 0.67 |

[a] The initial concentration was set at 1.0 for each plasmid and other concentrations are expressed relative to this value. Relative plasmid DNA concentration was unchanged after temperature upshift of an isogenic strain wild type for RNaseE.

## ANTISENSE RNA CONTROL OF IS*10* TRANSLATION

IS*10* is the mobile genetic element of the tetracycline resistance transposon Tn*10*. The IS*10*-encoded transposase (Tnp) is rate-limiting for both IS*10* and Tn*10* transposition and is regulated at a number of levels. One mechanism is antisense RNA control of *tnp* translation (Simons and Kleckner, 1983). IS*10* is 1329 *bp* and the *tnp* gene spans nearly the entire length (Fig. 4). The pIN promoter specifies the Tnp mRNA, termed RNA-IN. An opposing promoter, pOUT, specifies the antisense RNA, termed RNA-OUT. RNA-IN and RNA-OUT are complementary for 35 bp across a region that includes the Shine and Dalgarno sequence and AUG initiator codon of the *tnp* gene. RNA-OUT has a simple secondary structure in which a double-standed stem domain is topped by a loosely-structured loop domain (Fig. 5). The 5' end of the RNA-IN target is unstructured when the RNA is nascent. Genetic and physical studies show that RNA-OUT/RNA-IN pairing initiates by the formation of a few base pairs between the top of the RNA-OUT loop and the 5' nucleotides of RNA-IN, followed by rapid formation of the completely base-paired species as the stem structure is displaced (Case *et al.*, 1989; Kittle *et al.*, 1989; Fig. 5). The initial interactions are presumably reversible, although stable complex formation proceeds so rapidly that this equilibrium cannot be observed. Mutations (*e. g.* M1) that alter complementarity in the initiation region prevent pairing, whereas complementary changes in both RNAs in this region alter, essentially, only the specificity of the pairing reaction. In the RNA-OUT/RNA-IN reaction, the 5' end of RNA-IN must pass through the RNA-OUT loop in order to permit formation of double-stranded RNA prior to complete dissolution of base-pairing in the RNA-OUT stem. Thus, mutations that increase base-pairing in the RNA-OUT (*mcil*) or add additional sequences to RNA-IN (see Fig. 5) block pairing altogether. This model for RNA-OUT/RNA-IN pairing is in contrast to the several-step process of RNAI/RNAII pairing seen in plasmid ColE1 (*i. e.* kissing, followed by initiation of stable complex formation at the 5' end of RNAI; see above) or of CopA and CopT RNA pairing seen in plasmid R1 (see below).

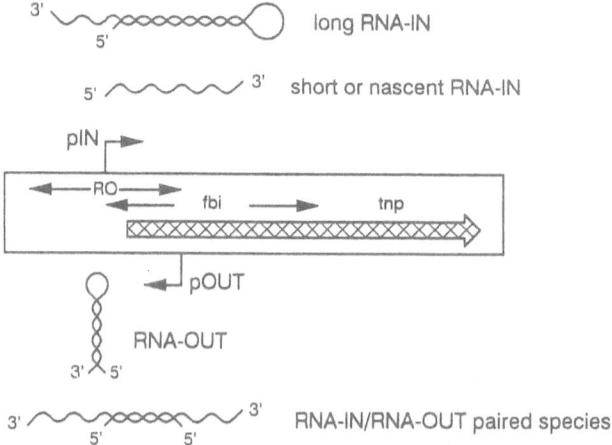

Fig. 4. The IS*10* element and its principal transcripts. The pIN and pOUT promoters specify, respectively, the *tnp* mRNA, RNA-IN, and the antisense RNA, RNA-OUT. The "RO" and "fbi" inverted repeats determine the secondary structure of RNA-OUT and RNA-IN, respectively. RNA-OUT pairs efficiently with short but not with long RNA-IN transcripts.

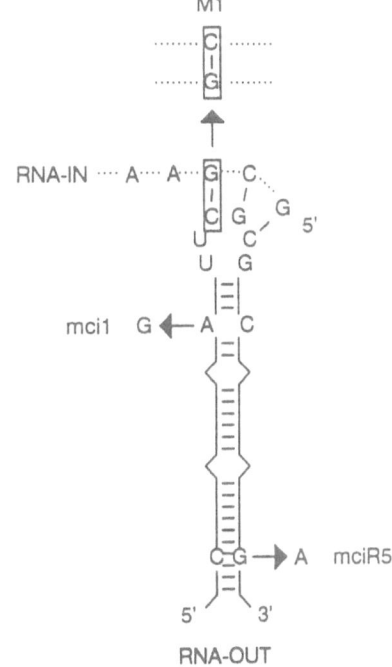

**Fig. 5.** Pairing between RNA-OUT and RNA-IN of IS*10*. The reaction initiates by the formation of a few base pairs between the RNA-OUT loop and the 5' end of RNA-IN. Mutations in this region (*e. g.* M1) prevent pairing when only one of the RNAs is changed, but alter pairing specificity when both RNAs are mutant. Mutations that stabilize the RNA-OUT loop domain(*e. g. mci*1) block pairing altogether. Mutations in the RNA-OUT stem domain (*e. g. mci*R5) alter RNA-OUT stability *in vivo*.

Mutations changing the RNA-OUT domain (*e. g. mci*R5) have no effect on the pairing betwen RNA-OUT and RNA-IN. Rather, they decrease RNA-OUT half life, which is normally quite long ($\approx$ 60 min at 37 °C). This high stability is due to the RNA-OUT stem domain; mutations that disrupt the stem structure destabilize RNA-OUT *in vivo*, but when a combination of such mutations restores stem structure, stability is also restored (Case *et al.*, 1989).

RNA-OUT inhibits *tnp* translation by directly blocking ribosome binding at the *tnp* translation-initiation site (Ma and Simons, 1990). The paired species is also cleaved efficiently by RNase III, which facilitates RNA-IN decay (Case *et al.*, 1990). However, such facilitated decay is not required for IS*10* antisense RNA control. The biological role of IS*10* antisense RNA control is probably to limit the accumulation of IS*10* elements in the cell; antisense RNA control works only when the IS*10* level rises to three or more copies per cell (J. Matsunaga and R. Simons, unpublished). The highly stable RNA-OUT is expressed at relatively low levels, in keeping with the very slow changes in IS*10* copy number. This arrangement is in contrast to rapidly changing plasmid copy number levels regulated by highly expressed, short-lived RNAs.

The structure of RNA-IN is also important for efficient control. RNA-OUT does not pair efficiently to RNA-IN transcripts longer than 215 nt. This is due to a downstream sequence in RNA-IN termed *fbi* (*f*old-*b*ack *i*nteraction), which is complementary to and sequesters the 5' end of RNA-IN, thereby preventing the initiation of pairing to RNA-OUT (Gonzalez and Simons, submitted; Fig. 4). Importantly, such mRNAs are not translated efficiently, due to sequestration of the *tnp* ribosome-binding site. Because only nascent mRNAs are translated, only these transcripts need to be inhibited by RNA-OUT. Thus, the overall effect of the *fbi* sequence is that it decreases the functional half life of RNA-IN and limits most *tnp* expression to nascent transcripts.

## ANTISENSE RNA CONTROL OF REPLICATION OF THE INCFII PLASMID R1

IncFII plasmids such as R1, R100 and R6-5 carry genetic information for three systems that are involved in plasmid maintenance and that are controlled by antisense RNA; these are vegetative replication, conjugal transfer and the Hok/Sok system that kills from within daughter cells that are borne without plasmid. In the present communication we will discuss the system that controls replication.

Plasmids control their own replication and they are present in defined copy numbers that are determined by the plasmid, the host and the growth conditions (Nordström, 1985). Hence, during steady-state growth, the frequency of replication is exactly one per plasmid copy and cell generation. If the copy number deviates from the steady-state value, the replication frequency is adjusted such that the steady-state copy number is maintained. Hence, plasmids are able to measure their concentration and to adjust their replication frequency accordingly. The same system is used for both these purposes and the control is negative. The control element is either a repressor protein, an antisense RNA, or DNA iterons (Nordström, 1985 and 1990). In most cases, the synthesis of a rate-limiting plasmid-encoded replication (Rep) protein is controlled. Exceptions are plasmids of the p15 and ColE1 families in which the processing of a preprimer transcript to a primer is controlled by antisense RNA (see above).

## Control of replication of plasmid R1

IncFII plasmids (*e. g.* R1) use the theta mode of replication. The smallest part of a plasmid that is still able to replicate with the normal copy number is called the basic replicon (Kollek *et al.*, 1978). The basic replicon of plasmid R1 and its control circuits are shown in Fig. 7.

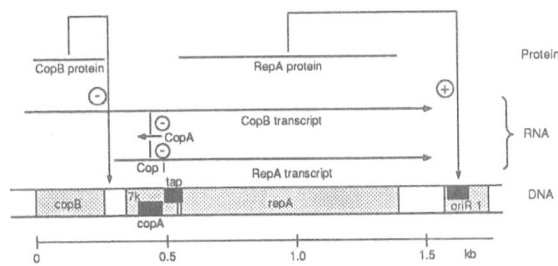

**Fig. 6.** Basic replicon of plasmid R1. Synthesis of the RepA protein is rate-limiting for replication and is negatively controlled transcriptionally by the CopB protein and post-transcriptionally by the CopA RNA.

Initiation of replication of plasmid R1 requires the binding of many molecules of the plasmid-encoded RepA protein to the origin (*ori*) of replication Masai *et al.*, 1983). The RepA protein is *cis*-acting (it preferentially binds to the origin on the DNA molecule from which its mRNA was formed). The *cis*-activity is caused by a region denoted CIS, located to the left of *ori* (Masai and Arai, 1988); deletion of CIS causes the RepA protein to function *in cis* as well as *in trans*. New RepA molecules have to be made for each round of replication, which makes the replication control very efficient. Replication is regulated by control of the rate of RepA synthesis (Nordström *et al.*, 1984). Two promoters (*pCopB* and *pRepA*) may be used to form the mRNA for the RepA protein. However, during normal conditions the *pRepA* promoter is almost totally switched off by the CopB protein (Light *et al.*, 1985). The *pCopB* promoter is constitutive and the control is exerted post-transcriptionally by an antisense RNA (CopA) that is encoded by a region about 80 bp upstream of the *repA* gene (Stougaard *et al.*, 1981).

The CopA RNA and its target (CopT, the upstream region of the RepA mRNA) form a duplex (Persson *et al.*, 1989) whose existence has been demonstrated *in vivo* by its cleavage by RNase III (Blomberg *et al.*, 1990). The interaction between CopA RNA and its target inhibits the expression of the RepA mRNA, which results in a negative correlation between the relative rate of plasmid replication (replications per plasmid copy) and the concentration (copy number) of the plasmid (Nordström 1990).

The key parameters in setting the plasmid copy number are *i*) the rates of synthesis and decay of RepA mRNA, *ii*) the intrinsic efficiency of RepA translation from RepA mRNA, *iii*) the concentration of CopA RNA, and *iv*) the kinetics of the interaction between CopA and CopT RNA.

**Table III**. Effect of the Rate of RepA Transcription of R1 Copy Number

| Replicon | Genotype *copA   copB* | | CopT Synthesis (Relative Scale) | Copy Number (Relative Scale) |
|---|---|---|---|---|
| Wild Type | *wt* | *wt* | 1.0 | 1 |
| Point mutation in the *copA* gene | *copA-104* | *wt* | 1.0 | 8 |
| Deletion of *pCopB* and part of the *copB* gene | *wt* | *copB⁻* | 2.0 | 8 |
| Point mutation in the *copB* gene | *wt* | *copB⁻* | 3.0 | 40 |

The *pCopA* promoter is constitutive and the CopA RNA is unstable with a $t_{1/2}$ of 1-2 min. Therefore, the concentration of CopA RNA is always proportional to the concentration of the plasmid. The pCopA promoter is 10-15 times stronger than the *pCopB* promoter. However, even moderate increases in RepA transcription drastically reduce the synthesis of CopA RNA due to convergent transcription (Nordström, 1992). Therefore, the copy number of plasmid R1 increases dramatically by even moderate increases in the rate of RepA transcription (Table III). This has been used to construct so-called runaway-replication plasmid vectors that can be amplified by small increses in temperature; amplification of the plasmid DNA occurs in the presence of protein synthesis (Uhlin *et al.*, 1979; Larsen *et al.*, 1984). Therefore, large quantities of protein may be obtained from a cloned gene (Nordström and Uhlin, 1992).

## Kinetics of duplex formation between CopA and CopT RNA

CopA and CopT RNA interact to form a duplex, a reaction that causes inhibition of the synthesis of RepA protein. Therefore, the kinetics of duplex formation are important for copy number control. In formulas and equations below, the symbols A and T are used for CopA and CopT, respectively, and their concentrations are denoted $A$ and $T$, respectively. The reaction between CopA and CopT RNA is as follows:

$$A + T \xrightarrow{\quad k_{app} \quad} AT \tag{1}$$

The rate constant, $k_{app}$, can be determined *in vitro* by a gel shift analysis and was found to be about $10^6$ l mole$^{-1}$ s$^{-1}$ (Persson *et al.*, 1989). Similar values have been found for duplex formation between RNAI and RNAII of plasmid ColE1 (Tomizawa, 1984) and RNA-In and RNA-OUT of IS*10* (Kittle *et al.*, 1989).

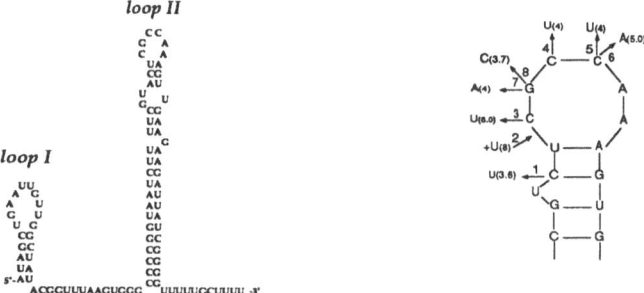

Fig. 7. Secondary structure of the antisense RNA (CopA) from plasmid R1 (left) (Wagner and Nordström, 1986) and the loop sequence of nine *copA* mutants (right). The figures within parenthesis show the relative increase in copy number caused by the mutations (*cf.* Nordström *et al.*, 1984).

Experimental studies of the secondary structure has shown that CopA RNA, as all natural antisense RNAs, has a high degree of secondary structure (Fig. 7)(Wagner and Nordström, 1986).

Genetic evidence suggested that the efficiency as well as the specificity of the interaction resides in loop II (Nordström *et al.*, 1984). Therefore, the rate constant $k_{app}$ was determined for the wild type and for *copA* mutants. The result was that a mutant whose copy number is 4-fold higher than that of the wild type showed a fourfold decrease in $k_{app}$ (Table IV)(Persson *et al.*, 1990b). It was therefore concluded that the copy number is determined kinetically by the rate of duplex formation. Furthermore, Brady *et al.* (1983) showed *in vivo* that the wild type was preferentially lost from cells carrying *copA* mutant 5 (cf. Fig. 7). This was verified by the inability of wild-type CopA to form a duplex with mutant CopT, whereas duplex formation was efficient in the opposite case (Table IV). Easton and Rownd (1982) reported that *copA* mutant 6 (*cf.* Fig. 7) is compatible with the wild-type R100; in this case, the heterologous pair would by A/G and C/U, respectively. Hence, the base-pair substitution totally changes the specificity of the interaction and the incompatibility group. These results indicate that the primary interaction involves the tip of loop II. On the other hand, *copA* mutant 3 (*cf.* Fig. 7) is also compatible with the wild type although the base-pair change is the same as that of mutant 5. This is probably due to the fact that the size of loop II of both CopA and CopT is reduced.

It may appear puzzling that a change in specificity of the CopA system may make a mutant plasmid compatible with the wild type since both plasmids still use the same rate-limiting protein. However, the explanation is that the RepA protein is *cis*-acting (Masai and Arai, 1988). In plasmid pT181, similar changes in the specificity of the control system have been reported, but they do not change the incompatibility properties because the RepC protein is acting equally effectively *in cis* and *in trans*. Consequently, the copy number of a wild-type plasmid in a heteroplasmid population increases and that of the mutant decreases due to sharing of the RepA molecules (Projan and Novick, 1984).

The importance of loop II as a determinant of the copy number prompted an *in vitro* analysis of the structural requirements for the reaction (Persson *et al.*, 1990a). Efficient duplex formation requires stem-loop II but also a sufficiently long (10-15 nt) single-stranded tail in either the 5' or the 3' end of stem II. A loop consisting of 6-7 nt is mostefficient in reaction (1)(Hjalt and Wagner, 1992). The wild-type sequence of loop II seems to be optimal, since all mutational changes examined so far lead to a reduced $k_{app}$.

**Table IV.** Kinetics of RNA duplex formation (Persson *et al.*, 1989). The nucleotide pair changed by the mutaiton is shown in bold letters.

| CopA | CopT | Hexanucleotides in loop II | Relative $k_{app}$ | Relative copy number |
|------|------|----------------------------|--------------------|----------------------|
| Wild type | Wild type | CGCCAA<br>GCGGUU | 1 | 1 |
| Mutant | Mutant | CGCUAA<br>GCGAUU | 0.2-0.3 | 3-4 |
| Mutant | Wild type | CGCUAA<br>GCGGUU | 0.2-0.3 | |
| Wild type | Mutant | CGCCAA<br>GCGAUU | <0.01 | |

A derivative of CopA that had lost most of the 5' tail is a much less efficient substrate for duplex formation (Persson *et al.*, 1990a). However, this truncated molecule turned out to be a competitive inhibitor of reaction (1). This opened the possibility to dissect the reaction between CopT and the truncated CopA (CopI): The two RNA species were allowed to interact. At intervals, the amount of free CopT was determined by its ability to rapidly form a duplex with added CopA. The reaction was found to proceed in at least two steps; in the first step a "kissing" intermediate (I:T) is formed (Persson *et al.*, 1990b). Then, the complete duplex is formed in a second step:

$$I + T \underset{k_{-1}}{\overset{k_1}{\rightleftharpoons}} I{:}T \xrightarrow{k_2} IT \qquad (2)$$

The kinetic constants were determined for the wild-type and *copA* mutant 5 (Table V). The rate constant $k_1$ was found to be about the same as the $k_{app}$ for the total reaction between CopA and CopT, $k_{-1}$ to be very low (about $10^{-5}$ s$^{-1}$), and $k_2$ to be about $10^{-4}$ s$^{-1}$. The kinetic analysis also showed that the *copA* mutation affected $k_1$ but did not change $k_2$. The $k_{-1}$ value for the mutant has not been determined yet.

The slow rate of the reaction from the kissing complex to complete duplex results in saturation kinetics; above a CopI concentration of about $3 \cdot 10^{-10}$ M, the rate of duplex formation is independent of the concentration of CopI (Persson *et al.*, 1990b) On the other hand, the reaction between CopA and CopT showed a linear correlation between the rate of the reaction and the concentration of CopA at least to a concentration of $10^{-7}$ M. The intracellular concentration of CopA has been estimated to be $\geq 10^{-7}$ M (Womble and Rownd, 1986). Hence, there is a linear correlation between the rate of duplex formation and the concentration of CopA in the range that is of biological significance (Persson *et al.*, 1990b). The rate-limiting step is the formation of the kissing complex, the step which is affected by *cop* mutations.

**Table V.** Kinetic constants of reaction (2) for wild type and *copA* mutant 5 (*cf.* Fig. 2)(Persson *et al.*, 1990b).

| CopI/CopT | $k_1$ (l mole$^{-1}$ s$^{-1}$) | $k_{-1}$ (s$^{-1}$) | $K_D = k_{-1}/k_1$ (M) | $k_2$ (s$^{-1}$) |
|---|---|---|---|---|
| Wild type | $3.4 \cdot 10^6$ | $\leq 10^{-5}$ | $10^{-11}$ | $1 \cdot 10^{-4}$ |
| Mutant | $0.6 \cdot 10^6$ | - | - | $1 \cdot 10^{-4}$ |

From the values of $k_1$ and $k_{-1}$, the dissociation constant $K_D$ was estimated to be about $3 \cdot 10^{-12}$ M. Hence, the complex is very stable (the $K_D$ corresponds to a $\Delta G$ value of about -100 BTU/mole) and it was possible to demonstrate the kissing complex between CopI and CopT on native gels. Furthermore, structural studies using single- and double-strand-specific nucleases indicated that the stable (deep-kissing) complex engages 6-8 base pairs (Persson *et al.*, 1990a).

The existence of a stable deep-kissing complex suggests that duplex formation proceeds in at least three steps with an unstable kissing complex (A:T, presumably engaging three loop nucleotides (GCC in CopA RNA)) and the deep-kissing complex (A::T) as intermediates.

$$A + T \underset{k_{-1}}{\overset{k_1}{\rightleftharpoons}} A{:}T \underset{k_{-2}}{\overset{k_2}{\rightleftharpoons}} A{::}T \overset{k_3}{\longrightarrow} AT \qquad (3)$$

The experimental measurements indicated that $k_1$ was affected by *copA* mutations. This may be somewhat misleading: large changes in $k_{-1}$ in reaction (3) would also lead to minor changes in the apparent rate of formation of the deep-kissing complex. Therefore, careful measurements of the rate of the backward reaction from A::T to A + T are badly needed but are difficult to perform because the rates are so low.

Measurements of $k_{app}$ as a function of temperature and salt concentration suggested that duplex formation proceeds gradually and that it does not involve an abrupt opening of the double-stranded stems before annealing to the complementary RNA species (Persson *et al.*, 1990a).

A scheme for RNA duplex formation was proposed that involves several steps and a gradual process once nucleation has started in the single-stranded regions at the bases of stem II (Fig. 8).

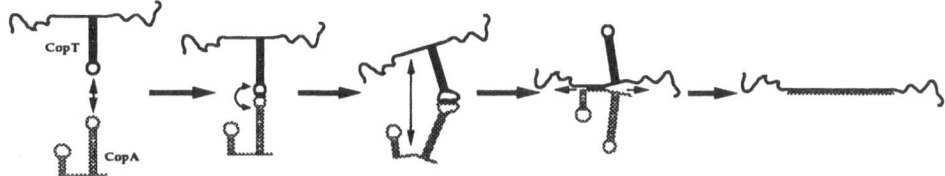

**Fig. 8**. Formation of the duplex between CopA and CopT RNA.

A stepwise process for duplex formation between an antisense RNA and its target was first proposed by Tomizawa (1984) for the interaction between RNAI and RNAII of plasmid ColE1. He found that there is an intermediate complex, the so-called kissing complex, formed by base-pairing betweeen the complementary loops known to be present in RNAI and RNAII. The reaction then proceeds by interaction between sequences outside of the stem loops. He was also able to determine the rate constants for the individual reactions. He further showed that the on-rate for the first reaction $(k_1)$ is rate-limiting for duplex formation and that the reaction leading from the kissing complex to complete duplex may be very rapid. The reaction between RNA-IN and RNA-OUT of IS*10* is processive starting with a reaction between the 5' end of RNA-IN and the loop structure of RNA-OUT. No discrete intermediates have been identified.

Antisense RNAs and/or their targets contain bulges in their stems, particularly in the upper parts of the stems (*cf.* Fig. 7). This might facilitate the opening of the structure and thereby the formation of the deep-kissing complex, a conclusion that is supported by experimental evidence that show that $k_{app}$ is reduced by about 20-fold when the bulges are removed and sealed by mutation. However, another, maybe more important consequence of the presence of the bulges is that they render the molecule resistant to cleavage by RNase III; RNase III reduces the concentration of a bulge-free CopA in the cells to very low *in vivo* levels (Hjalt and Wagner, unpublished).

The kinetic experiments were performed *in vitro* in the absence of protein. There is no evidence (genetic or other) of the existence of any protein that is specifically involved or required for duplex formation betweeen CopA and CopT RNA *in vivo*.

If the kinetics were exactly the same *in vivo* as *in vitro*, the value of $k_1$ (about $10^6$ 1 mole$^{-1}$ s$^{-1}$) and the estimated concentration of CopA ($\geq$ $10^{-7}$ M) gives:

$$v = k_1 \cdot A \cdot T \geq 10^{-1} \cdot T$$

*i. e.*, the half life of free T is $\leq 10$ s. This makes the reaction rapid enough to be of biological significance.

Determinations of copy number of mutants and experiments with *repA-lacZ* fusions have shown that

$$dR1/dt \sim dRepA/dt$$

The output of RepA protein is determined by the rate of synthesis of RepA-mRNA, the intrinsic translational efficiency of RepA-mRNA ($k_a$), the concentration of CopA RNA, and the rate of duplex formation between CopA and CopT RNA ($k_{app}$). The rate of synthesis of RepA-mRNA is determined by the strength of the promoter and the concentration of the plasmid ($R1$); the concentration of CopA RNA is determined by the stregth of the *pCopA* promoter, the concentration of the plasmid and the half life of CopA RNA. This can be summarized as follows:

$$dR1/dt \sim dRepA/dt = (k_a \cdot dRepA\text{-}mRNA/dt)/k_c \cdot [CopA] \cdot k_{app} =$$
$$= K \cdot (dRepA\text{-}mRNA/dt)/[CopA] \cdot k_{app} =$$
$$= K_1 \cdot pRepA \cdot [R1]/pCopA \cdot [R1] \cdot (1/t_{1/2}) \cdot k_{app} = Constant \qquad (4)$$

As is shown by eq. (4), the rate of replication is predicted to be independent of the concentration of the plasmid, the $+n$ mode of replication (Nordström and Aagaard-Hansen, 1984), and experiments with copy-number shifts support this conclusion (Gustafsson and Nordström, 1980).

## Translational inhibition caused by the interaction between CopA and CopT RNA

The interaction between CopA and CopT RNA results in a total inhibition of RepA synthesis. The end point of the RNA duplex is 80 nt upstream of the RepA start codon. This suggests that the coupling between duplex formation and translational inhibition is indirect. Several molecular explanations for this indirect effect have been proposed, *i*) changed secondary structure of the RepA mRNA, *ii*) changed half life of RepA-mRNA, and *iii*) premature transcriptional attenuation/termination. In the case of plasmid R1, all three explanations have been rejected:

Dong *et al.* (1987) performed computer predictions of the secondary structure of the RepA mRNA in the absence and in the presence of duplex formation. They predicted that the Shine-Dalgarno region and the start codon of the RepA message changed from single- to double-strandedness as a consequence of binding of CopA RNA to CopT. However, *in vitro* studies as well as mutational analyses did not support this model (Berzal-Herranz *et al.*, 1991; Öhman and Wagner, 1991). Direct *in vitro* measurements did not show any significant decrease in the half life of the RepA message due to RNA duplex formation (Blomberg and Wagner, unpublished). Finally, no evidence was found that supported premature attenuation/termination of RepA transcription in the presence of CopA RNA (Blomberg and Wagner, unpublished). Premature transcriptional termination induced by the binding of the antisense RNA is used by pT181, a staphylococcal plasmid, to control expression of the rate-limiting protein, RepC (Novick *et al.*, 1989).

RNase III cleaves the RNA duplex. The expression of a *repA-lacZ* fusion gene was found to be greatly increased in an *rnc* mutant. Similarly, a copy mutant (the *copB* deletion mutant shown in Table III) could not be established in the *rnc* host (Blomberg *et al.*, 1990). However, the wild-type R1 as well as the *copA-104* mutant (*cf.* Table III; mutant 3 in Fig. 7) could establish and were present in the normal copy number in the *rnc* host. The reason for these discrepancies is at present unclear.

Recently, an open reading frame consisting of 24 codons was discovered in the region beween the *copA* and *repA* genes (see *tap* in Fig. 7). This open reading frame is expressed and its expression is inhibited by CopA RNA to the same extent as is expression of the *repA* gene (Blomberg *et al.*, 1992). A puzzling finding is that kissing is enough to cause inhibition of Tap synthesis (Wagner *et al.*, 1992).

A mutational analysis showed that translation of the *tap* gene is necessary for synthesis of the RepA protein (Blomberg *et al.*, 1992). Translation of the reading frame (*tap*, *t*ranslational *a*ctivator *p*rotein) rather than the protein itself is needed for RepA synthesis. The RepA translation initiation region is inactive, maybe because of sequestering in a stable stem-loop strucure or to suboptimal spacing of the Shine-Dalgarno and the initiation codon. Presumably, translation of the *tap* message, which ends in the second codon of the RepA message, opens up this stem and allows the ribosome to restart at the RepA start codon. When the Tap stop codon was moved 19 codons further into the RepA message, virtually no RepA protein was formed, indicating that the translational coupling requires termination of Tap translation close to the start signals of the RepA reading frame. A model for control of RepA synthesis by translational coupling is shown in Fig. 9.

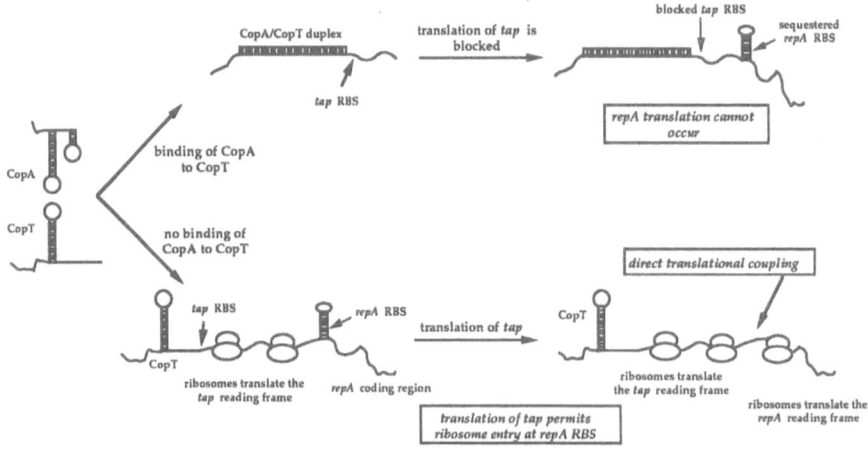

**Fig. 9.** Control of RepA synthesis by antisense RNA and translational coupling between the synthesis of the leader peptide and of the RepA protein.

Several unrelated plasmids use translational coupling to control the synthesis of a rate-limiting Rep protein. A well studied case is plasmid ColIb-P9 (Hama *et al.*, 1990). Also in this case, translation of a leader peptide, RepY, is necessary for opening of the start region of the RepZ message. The on-state is stabilized by the formation of a pseudoknot between the CopT loop and part of the stem structure present in the vicinity of the RepZ start. Binding of the antisense RNA to the CopT loop inhibits the formation of the pseudoknot and RepZ synthesis is no longer possible. Results obtained in the R1 system appear to exclude the involvement of a similar pseudoknot.

What is the significance of the presence of genes for leader peptides? Why isn't the control exerted directly on the Rep message? The most obvious explanation is that it would be difficult to move the Rep message closer to the region where the RNA duplex is located, because the promoter for the antisense RNA would then be situated within the structural gene for the Rep protein. This would presumably restrict the choice of amino acids in the Rep protein. Another, perhaps more likely, possibility is that the system allows a very low rate of synthesis of the RepA protein; the Tap reading frame and reinitiation at the RepA start would the act as a gear box.

## Concluding remarks

Plasmid R1 uses a very efficient system to measure its copy number and to adjust its replication frequency in order to maintain its steady state copy number. The concentration of an antisense RNA (CopA) is used as a measure of the concentration of the plasmid and the key reaction in copy number control is an interaction between the antisense RNA and its target, an upstream region of the RepA message. This interaction is a second-order reaction that takes place in several steps. The rate-limiting event is the first reaction that leads to the formation of a kissing intermediate that is very stable. The kinetics of the system gives a linear correlation between the rate of RepA syntesis and the inverse value of the concentration of CopA RNA. Hence, the system responds efficiently to changes in copy number and the rate of replication is independent of the copy number of the plasmid (the $+n$ mode of replication). The interaction between CopA and CopT RNA leads to the formation of an RNA duplex, which results in inhibition of the synthesis of a leader peptide (Tap); translational coupling to the synthesis of Tap is required for the synthesis of the RepA protein.

## NATURAL ANTISENSE RNA CONTROL IN EUKARYOTES

In recent years, antisense transcription, that is, transcription of both strands of a single DNA segment, has been documented in a variety of eukaryotic cells (Simons, 1993; see Table VI). Unlike the prokaryotic cases, however, none of these examples conclusively demonstrates natural antisense RNA control. Nevertheless, circumstantial evidence strongly suggests that such control does operate in eukaryotic cells. This evidence includes the conservation of antisense transcription patterns, demonstration of duplex formation *in vivo*, evidence of double-stranded RNA unwindase modification, detection of complementary sense and antisense RNAs in the same cell or tissue, and correlations between antisense RNA production and target RNA activity. Indirect evidence comes from the many cases of effective control by artificial antisense RNAs.

Antisense could regulate eukaryotic gene expression in many ways, including attenuation of transcription, regulation of nuclear RNA splicing, inhibition of RNA translocation to the cytoplasm, inhibition of translation, alteration in nuclear or cytoplasmic RNA stability, and the co-translocation and/or co-translation of related functions encoded on complementary mRNAs. Here we briefly describe three cases where the evidence for eukaryotic antisense RNA control is reasonably strong.

The most often cited and perhaps best example is that of the *Xenopus* basic fibroblast growth factor (bFGF), a potent mitogen that converts ectoderm to mesoderm prior to embryonic gastrulation. The bFGF mRNA is present in the oocyte but disappears soon after fertilization, not to reappear until neural tube closing. Presumably, it is important that bFGF is expressed in the oocyte but not in the early embryo. The bFGF region also specifies a processed and polyadenylated antisense transcript that is complementary to exon 3 of the bFGF mRNA. The bFGF mRNA is degraded soon after germinal vesicle breakdown, at which time about 50% of the mRNA adenines in the complementary region are deaminated to inosine (Kimelman and Kirschner, 1989). Such RNA modification is very typical of the double-stranded RNA deaminase (unwindase) activity present in *Xenopus* embryos and other cells (Nishikura, 1992), strongly suggesting that the bFGF and antisense RNAs are paired at the time of mRNA decay. Thus, antisense RNA is implicated in *bFGF* control.

Several protooncogenes specify antisense transcripts, but only *N-myc* antisense RNA (termed *N-cym*) is known to be stable. The best evidence implicating *N-cym* RNAs in *N-myc* control comes from studies showing that *N-myc/N-cym* RNA duplexes very likely form *in vivo*

(Krystal *et al.*, 1990). Equimolar amounts of RNase A-resistant *N-myc* and *N-cym* RNA species are recovered immediately after cell lysis, unless the extract is first heated. These and related observations strongly suggest that the duplexes form within the cell. These studies also show that *N-cym* RNAs duplex preferentially with *N-myc* precursor RNAs that retain intron 1, suggesting that *N-cym* modulates *N-myc* splicing.

**Table VI**. Some examples of antisense transcription in eukaryotic cells[a)]

Conserved patterns of antisense genes and transcription
  *Xenopus*, human and cow *bFGF*; avian, human and rat *erbA*; yeast
  RAD10 and human ERCC-1; rat and human GnRH; mammalian
  *Surf*; mouse and hamster *dfhr*
Antisense transcription of proto-oncogenes
  Human *N-myc* and *L-myc*; mouse *c-myc* and *c-myb*
Antisense transcription of short repeated sequences
  B1 and B2 elements in mouse p53 and αFP genes
Transposable elements in which antisense RNAs derive from flanking
  promoters
  *Chlamydomonas* TOC1 and *Drosophila copia* elements
Antisense RNAs resulting from gene rearrangements
  Inversion/duplication of the mouse *mid* gene; "flipped"
  pseudogene of murine L27' gene
Intronic genes transcribed in antisense direction
  *Drosophila PCP* gene in *Gart* and *Pig-1* gene in *dunce*; antisense
  genes in human monkey, rat and mouse factor VIII and human
  *NF1LT* genes
Viral cases
  Herpes simplex LAT RNA; Epstein-Barr antisense transcripts;
  human papilloma virus
Other cases
  Avian tcRNA; yeast *PET122*, *HAP3* and *CDC9/CDC36*; β-casein;
  chloroplast *pcbB*; *Drosophila Dcd*

a) For key references, see Simons 1993

Herpes simplex virus establishes latent infections in the neurons of sensory ganglia, from which reactivation can occur. The *ICP0* gene encodes a potent transactivator of viral gene expression and has been implicated in reactivation from latency. The *LAT* RNA, which is antisense to the 3' region of the *ICP0* RNA, is a stable, polyadenylated intron deriving from the same region (Stevens *et al.*, 1987; Farrell *et al.*, 1991). Recent genetic experiments show that when *LAT* RNA is expressed *in trans* from a constitutive promoter, *ICP0* expression is inhibited. Despite this clear evidence of *LAT* RNA function, its role in the viral life cycle must be quite subtle; *LAT* RNA is apparently dispensable for viral growth as well as for the establishment, maintenance, and reactivation of the latent state.

## ARTIFICIAL ANTISENSE RNA CONTROL

Fairly efficient and specific artificial antisense RNA control has been demonstrated in a number of different organisms (Mol and van der Krol, 1991; Erickson and Izant, 1992). Generally, such control is engineered by transcribing a copy of the gene of interest in the antisense direction. Antisense transcripts can be expressed *in vivo* from an appropriate vector or, in some cases, injected into the cell or embryo of interest following synthesis *in vitro*. In either case, very high concentrations of antisense transcripts are usually necessary to achieve a substantial level of control. This general approach to creating artificial "mutations" has identified the function of a number of genes and is particularly useful in organisms that lack well developed or rapid classical genetic approaches.

While artificial antisense transcripts have proved to be powerful experimental tools, obvious delivery problems have slowed their application for therapeutic purposes. However, antisense oligodeoxy-ribonucleotides (ODNs) hold considerable promise in this regard and are currently the subject of intensive drug discovery and development efforts.

Natural antisense RNA control should provide lessons for the design of efficient and specific artificial RNA or ODNs. Chief among these lessons are those learned form the defined pairing mechanisms (see above). If a known or putative stem/loop structure forms at an appropriate position in the target RNA, an antisense RNA or ODN can be rationally designed to follow either a one- (*cf.* Fig. 4) or a two-step (*cf.* Fig. 8) pathway. For an effective one-step scheme, the antisense molecule should be complementary to one side (5' or 3') of the target structure and not possess significant amounts of self complementarity. For an effective two-step process, the antisense should contain a stem/loop structure complementary to that of the target, plus an unstructured 5' and/or 3' tail. These and other obvious designs for efficient artificial antisense RNAs have been or are currently being investigated (Persson *et al.*, 1990a; B. Zeiler and R. Simons, unpublished; F. Söderbom and E. G. H. Wagner, unpublished).

## ADVANTAGES OF ANTISENSE RNA CONTROL

The advantages of control by antisense RNA compared to a protein repressor are not yet clear. Certainly, it can be argued that some level of antisense RNA control of a gene might be achieved by the evolution of an appropriately positioned, opposing promoter. Such evolution seems more easily achieved than that of a nucleic acid binding protein. Indeed, antisense RNA control might be primordial, possibly predating the evolution of functional proteins. On the other hand, the mere production of a complementary transcript is not sufficient to establish efficent and specific control. Rather, efficient control requires specific sequences and structures in both the antisense and target RNAs (see above).

Antisense RNAs vary considerably in their metabolic stability. In the control of plasmid replication, a relatively unstable antisense RNA is expressed constitutively. Thus, the concentration of the antisense RNA will always be closely proportional to the copy number of the plasmid. Furthermore, because antisense and target RNA pairing is a second order reaction, the rate of pairing will be proportional to the concentration of the antisense RNA. Thus, over a potentially broad range, plasmid concentration (copy number) can be held quite stable. Control by an unstable protein repressor would achieve the same goal, but it is far more costly for the cell to produce protein than RNA.

For those systems in which the antisense and target RNAs are specified by opposing promoters, mutations in regions critical for the pairing interaction will, in many cases, alter pairing specificity rather than abolish control altogether (see above). Thus, a distinct advantage of an antisense RNA system may be its ability to rapidly evolve new specificity groups by single point mutations.

## Acknowledgements

The work presented was supported by grants from the Swedish Cancer society (K. N.), The Swedish Natural Science Research Council (K. N.), The U. S. National Institute of General Medical Sciences (Grant GM 27241 to S. N. C.), The U. S. Public Health Service (R. W. S.) ), and The American Cancer Society (R. W. S.).

## REFERENCES

Berzal-Herranz A, Wagner EGH and Diaz-Orejas R (1991) Control of replication of plasmid R1: The intergenic region between copA and repA modulates the level of expression of repA. Mol Microbiol 5:97-108

Blomberg P, Wagner EGH and Nordström K (1990) Control of replication of plasmid R1: The duplex between the antisense RNA, CopA, and its target, CopT, is processed specifically *in vivo* and *in vitro* by RNaseIII. EMBO J 9:2331-2340

Blomberg P, Wagner EGH and Nordström K (1992) Replication control, of plasmid R1: RepA synthesis is regulated by CopA RNA through inhibition of leader peptide translation. EMBO J 11:2675-2683

Brady G, Frey J, Danbara H and Timmis KN (1983) Replication control mutations of plasmid R6-5 and their effects on interactions of the RNA-F control element and its target. J Bacteriol 154:429-436

Case CC, Roels SM, Jaensen PD, Lee J, Kleckner N and Simons RW (1989) The unusual stability of the IS10 anti-sense RNA is critical for its function and is determined by the structure of its stem-domain. EMBO J 8:4297-4305

Case CC, Simons EL and Simons RW (1990) The IS10 transposase mRNA is destabilized during antisense RNA control. EMBO J 9:1259-1266

Cesareni G and Banner DW (1985) Regulation of plasmid copy number by complementary RNAs. Trends Biochem Sci 10:303-306

Dong X, Womble DD and Rownd RH (1987). Transcriptional pausing in a region important for plasmid NR1 replication control. J Bacteriol 169:5353-5363

Easton AM and Rownd RH (1982) The incompatibility product of IncFII plasmid NR1 controls gene expression in the plasmid replication region. J Bacteriol 152:829-839

Eguchi Y, Itoh T and Tomizawa J-I (1991) Antisense RNA. Annu Rev Biochem 60:631-652

Erickson RP and Izant JG (1992) Gene Regulation: Biology of Antisense RNA and DNA, Raven Press, Ltd, New York

Farrell MJ, Dobson AT and Feldman LT (1991) Herpes simplex virus latency-associated transcript is a stable intron. Proc Natl Acad Sci USA 88:790-794

Fitzwater T, Zhang X-Y, Elble R And Polisky B (1988) Conditional high copy number ColE1 mutants: resistance to RNAI inhibition *in vivo* and *in vitro*. EMBO J 7:3289-3297

Gonzalez JE and Simons RW (1993) Antisense RNA control of IS10 transposition: a site within the transposase messenger RNA blocks antisense pairing. Submitted

Gonzalez JE and Simons RW (1993) Translation of the IS10 transposase: messenger RNA conformation determines ribosome binding. Submitted

Gustafsson P and Nordström K (1980) Control of replication of plasmid R1. Kinetics of replication in shifts between different copy number levels. J Bacteriol 141:106-110

Hama, C, Takizawa T, Moriwaki H and Mizubuchi K (1990) Role of leader peptide synthesis in *repZ* gene expression of the CollIb-P9 plasmid. J Biol Chem 265:10666-10673

Hjalt T and Wagner EGH (1992) The effect of loop size in antisense and target RNAs on the efficiency of antisense RNA control. Nucl Acids Res 20:6723-6732

Kimelman D and Kirschner MW (1989) An antisense mRNA directs the covalent modification of the transcript encoding fibroblast growth factor in Xenopus oocytes. Cell 59:687-696

Kittle JD, Simons RW, Lee J and Kleckner N (1989) Insertion sequence IS10 anti-sense pairing initiates by an interaction between the 5' end of the target RNA and a loop in the anti-sense RNA. J Mol Biol 210:561-572

Kollek R, Oertel W and Goebel W (1978) Isolation and characterization of the minimal fragment required for autonomous replication of a copy mutant (pKN102) of the antibiotic resistance factor R1. Mol Gen Genet 162:51-58

Krinke L and Wulff DL (1987) OOP RNA, produced from multicopy plasmids inhibit λ cII gene expression through an RNase III-dependent mechanism. Genes Dev 1:1005-1013

Krystal GW, Armstrong BC and Battey JF (1990) N-myc mRNA forms an RNA-RNA duplex with endogenous antisense transcripts. Mol Cell Biol 10:4180-4191

Larsen JEL, Gerdes K, Light J and Molin S (1984) Low-copy-number plasmid-cloning vectors amplifiable by derepression of an inserted foreign promoter. Gene 28:45-54

Light J, Riise E and Molin S (1985) Transcription and its regulation in the basic replicon region of plasmid R1. Mol Gen Genet 198:503-508

Lin-Chao S, Chen W-T and Wong T-T (1992) The high copy number of pUC plasmid results from a Rom/Rop suppressible point mutation in RNAII. Mol Microbiol 6:3385-3393

Lin-Chao S and Cohen SN (1991) The rate of processing and degradation of antisense RNAI regulates the replication of ColE1 type plasmids in vivo. Cell 65:1233-1242

Ma C and Simons RW (1990) The IS10 antisense RNA blocks ribosome binding at the transposase translation initiation site. EMBO J 9:1267-1274

Masai H, Kaziro Y and Arai, K-I (1983) Definition of oriR, the minimum DNA segment essential for initiation of R1 plasmid replication *in vitro*. Proc Natl Acad Sci USA 80:6814-6818

Masai H and Arai K-I (1988) *RepA* protein- and *oriR*-dependent initiation of R1 plasmid replication: identification of a *rho*-dependent transcription terminator required for *cis*-action of *repA* protein. Nucl Acids Res 84:6493-6514

Mol JNM and van der Krol AR (1991) Antisense Nucleic Acids and Proteins, Marcel Decker, Inc, New York.

Nishikura K (1992) A cellular activity that modifies and alters the structure of double-stranded RNA. In Gene Regulation: Biology of Antisense RNA and DNA. (RP Ericksen and JG Izant, eds), Raven Press, New York, pp 21-34

Nordström K (1985) Control of plasmid replication - Theoretical considerations and practical solutions. pp. 189-214 In Plasmids in Bacteria (D Helinski et al., eds.), Plenum, pp 189-214

Nordström K (1990) Control of plasmid replication -How do DNA iterons set the replication frequency? Cell 63:1121-1124

Nordström K and Uhlin BE (1992) Runaway-replication plasmids as tools to produce large quantities of proteins from cloned genes. Bio/Technology 10:661-666

Nordström K and Aagaard-Hansen H (1984) Maintenance of bacterial plasmids: Comparison of theoretical calculations and experiments with plasmid R1. Mol Gen Genet 197:1-7

Nordström K, Molin S and Light J (1984) Control of replication of bacterial plasmids: Genetics, molecular biology, and physiology of the plasmid R1 system. Plasmid 12:71-90

Novick RP (1987) Plasmid incompatibility. Microbiol Rev 51:381-395

Novick RP, Iordanescu S, Projan S, Kornblum J and Edelman I (1989) pT181 plasmid replication is regulated by a countertranscript-driven transcriptional attenuator. Cell 59:395-404

Öhman M and Wagner EGH (1991) Regulation of replication of plasmid R1: An analysis of the intergenic region between copA and repA. Mol Gen Genet 230:321-328

Persson C, Wagner EGH and Nordström K (1989) Control of replication of plasmid R1: Kinetics of in vivo interaction between the antisense RNA, CopA, and its target, CopT. EMBO J 7:3279-3288

Persson C, Wagner EGH and Nordström K (1990a) Control of replication of plasmid R1: Structures an sequences of the antisense RNA, CopA, required for its binding to the target RNA, CopT. EMBO J 9:3767-3775

Persson C, Wagner EGH and Nordström K (1990b) Control of replication of plasmid R1: Formation of an initial transient complex is rate-limiting for antisense RNA/target RNA pairing EMBO J 9:3777-3785

Polisky B (1988) ColE1 replication control circuitry: sense from antisense. Cell 55:929-932

Projan SJ and Novick RP (1984) Reciprocal intrapool variation in plasmid copy numbers: A characteristic of segregational incompatibility. Plasmid 12:52-60

Simons RW (1991) Natural antisense RNA control in bacteria, phage and plasmids. In Antisense Nucleic Acids and Proteins (JNM Mol and AR van der Krol, eds), Marcel Decker, Inc, New York, pp 7-45

Simons RW (1993) The control of prokaryotic and eukaryotic gene expression by naturally occuring antisense RNA. In Antisense Research and Applications (ST Crooke and B Lebleu, eds) CRS Press, Inc, Boca Raton, Florida, in press

Simons RW and Kleckner N (1983) Translational control of IS10 transposition. Cell 34:683-691

Simons RW and Kleckner N (1988) Biological regulation by antisense RNA in prokaryotes. Annu Rev Genet 22:567-600

Stevens JG, Wagner EK, Devi-Rao GB, Cook ML and Feldman LT (1987) RNA complementary to a herpes virus alpha gene mRNA is prominent in latently infected neurons. Science 235:1056-1059

Stougaard P, Molin S and Nordström K (1981) RNA molecules involved in copy number control and incompatibility of plasmid R1. Proc Natl Acad Sci USA 78:6008-6012

Tomcsdanyi T and Apirion D (1985) Processing enzyme ribonuclease E specifically cleaves RNAI an inhibitor of primer formation in plasmid DNA synthesis. J Mol Biol 185:713-720

Tomizawa J (1984) Control of ColE1 plasmid replication: the process of binding of RNAI to the primer transcript. Cell 38:861-870

Tomizawa J and Som T (1984) Control of ColE1 plasmid replication: enhancement of binding of RNAI to the primer transcript by the Rom protein. Cell 38:871-878

Uhlin BE, Molin S, Gustafsson P and Nordström K (1979) Plasmid cloning vehicles with temperature-dependent copy number control: Amplification of genes and gene products. Gene 6:91-106

Wagner EGH and Nordström K (1986) Structural analysis of an RNA involved in replication control of plasmid R1. Nucl Acids Res 14:2523-2538

Wagner EGH, Blomberg P and Nordström K (1992) Replication control in plasmid R1: Duplex formation between antisense RNA, CopA, and its target, CopT, is not required for inhibition of RepA synthesis. EMBO J 11:1195-1203

Womble DD and Rownd RH (1986) Regulation of IncFII plasmid DNA replication. A quantitative model for control of plasmid NR1 replication in the bacterial cell division cycle. J Mol Biol 192:529-48

# Subject Index

# NATO ASI Series H

# NATO ASI Series H

# NATO ASI Series H

# NATO ASI Series H

# NATO ASI Series H